MOLECULAR MODELLING
AND
DRUG DESIGN

MOLECULAR MODELLING
AND
DRUG DESIGN

Dr. K Anand Solomon
Assistant Professor
Department of Bioinformatics
Sri Ramachandra University
Chennai

MJP PUBLISHERS

Cataloguing-in-Publication Data

Anand Solomon, K (1973 –).
  Molecular Modelling and Drug Design / by K. Anand Solomon. –
Chennai : MJP Publishers, 2008
  xii, 226p. ; 23 cm.
  Includes Glossary, References and Index.
  ISBN 978-81-8094-060-6 (pbk.)
  1. Biology, Molecular-Modelling
2. Drug Design, 3. Modelling, Molecular Biology. I. Title.
  572.8:615.19 ANA MJP 048

ISBN 978-81-8094-060-6
© Publishers, 2008
All rights reserved
Printed and bound in India

MJP PUBLISHERS
47, Nallathambi Street
Triplicane
Chennai 600 005

Publisher : J.C. Pillai
Managing Editor : C. Sajeesh Kumar
Project Editor : P. Parvath Radha
Acquisitions Editor : C. Janarthanan
Assistant Editor : B. Ramalakshmi
Composition : N. Yamuna Devi, Lissy John
Cover Designer : N. Yamuna Devi
CIP Data : Prof. K. Hariharan

This book has been published in good faith that the work of the author is original. All efforts have been taken to make the material error-free. However, the author and publisher disclaim responsibility for any inadvertent errors.

Preface

Computer assisted calculation has integrated itself as an important aspect in the field of scientific research. This is more so in the integrated disciplines such as biophysics, biochemistry, nanotechnology, etc. Computers simplify the calculations that might take years to complete through manual effort. One notable aspect of such computation is their predictive and investigative nature. Be it computational biology/chemistry/toxicology, the models derived from such calculations throw light in understanding, say, the characteristic nature of a product in a chemical reaction or how a protein folds, eliciting a diseased condition in the human body. Impressive progress has been made in such directions in the field of 'Drug Design', termed as *in silico* calculations. All drugs are chemicals, but the reverse is not true. How to characterize a chemical with drug-like properties? How do we predict the efficacy of a chemical to function as a drug? Is it possible to predict the optimum time a drug needs to be in the body to effect its action? Apart from its beneficial action as a drug, is it possible to predict its toxicity—commonly termed as 'side effects'? All these queries, no doubt can be addressed by wet-lab experimental studies (*in vivo/in vitro*), but also adds on to the time and human resources involved in it.

In contrast, *in silico* studies consume less of these and can be done in the absence of animal models and laborious procedures. *Molecular Modelling* is an aspect of such studies, wherein, say a protein/drug/their complex is modelled and studied under different conditions. Conceptually an attempt is made to mimic a process that occurs in the human system and study it in detail to get a better understanding.

This book is an introduction to the field of molecular modelling and drug design. An attempt has been made to elucidate the process with case studies.

I am very grateful to Professor S.S. Rajan, Head, Department of Crystallography and Biophysics, who has been my research mentor. To him I owe the knowledge that I have imbibed on Crystallography and other such aspects.

I wish to place my special thanks to Ms. Hemalatha, Lecturer, Department of Bioinformatics, Sri Ramachandra University, for going through the material and offering me positive criticism for improving the presentation of this book.

I also wish to thank the Management of Sri Ramachandra University for encouraging me in writing this book.

To my parents Mr. Kamalakaran and Mrs. Rajalakshmi, I owe a very loving gratitude, because, without them, I would not have been whatever little Iam. To my better half, Janani, I extend my love and acknowledgement for being supportive in all my efforts.

I wish to commend and place a special note of thanks to MJP Publishers, for their effective work in editing and formatting this book and for their friendly interactions.

K. Anand Solomon

Contents

1

MODELLING THE MOLECULES AND DESIGNING THE DRUGS

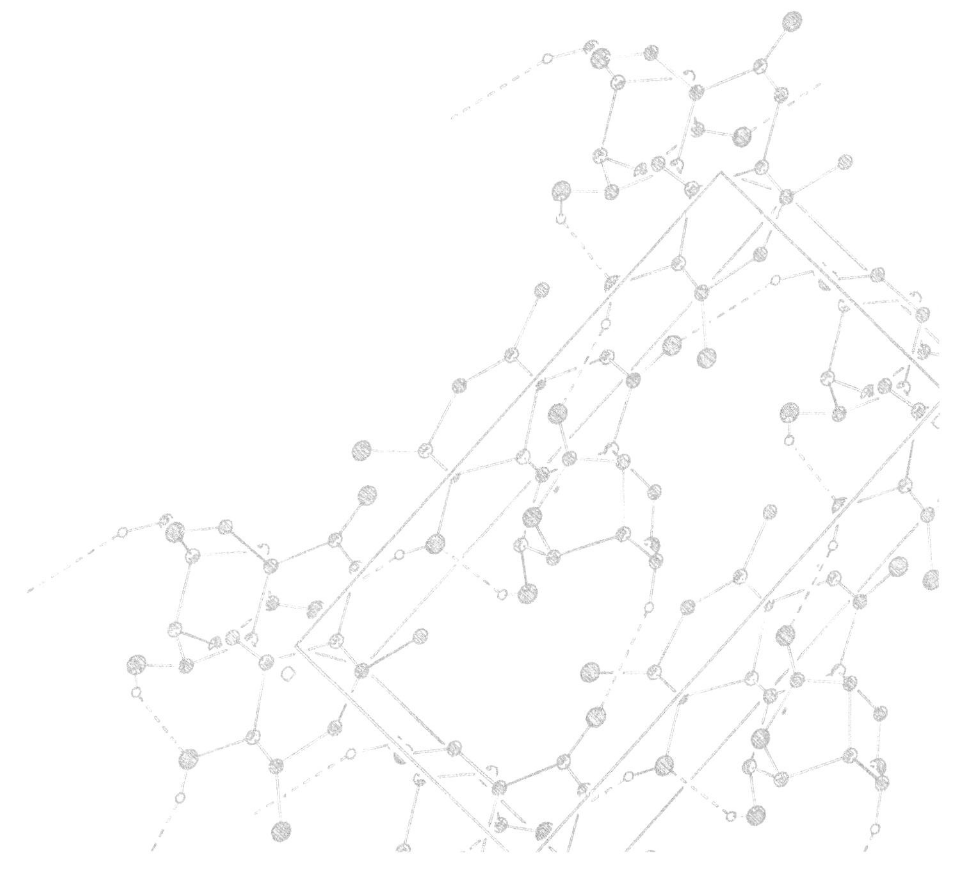

INTRODUCTION

Evolution of science has been benefiting mankind through a myriad ways, the notable being in tracking and controlling diseases. It is very difficult to imagine a world devoid of medicinal drugs. Drugs, being chemical, have their relative side effects in spite of which they are essential in dealing with life and disease. Diseases are caused due to the disharmony in the bodily parts (internal/external) wherein they either overdo or under-do their work. To treat and control a disease, it is important to understand the biological processes involved in its evolution. Be it a simple headache or life-threatening cancer, all involve some biological processes like cell-to-cell communication, neural transmission, etc. Such processes in the human system are very complex, and evaluating them would not only be time-consuming and tedious but also give only inaccurate results. The better way would be, to simulate the biological processes, and understand and design methodology for tackling the disease. *In vivo* and *in vitro* studies constitute the experiments for simulating them in the wetlab whereas *in silico* (computer aided) methods do not need animal models or enzymatic methods. *In silico* approaches have gained immense popularity and have become an integral part of the industrial and academic research that is directed towards drug design and discovery.[1-4]

Computational scientists combine their knowledge of molecular interactions and drug activity, together with visualization techniques, detailed energy calculations, geometric considerations, and data filtered out of huge databases, in an effort to narrow down the search for effective drugs.

In those cases wherein experimental evidence/data are not available, it is necessary to have some models to work on the *in silico* bench. Models can be structures drawn *in silico* and resemble the experimental structure to as great an extent as possible. For example, in order to build a chemical model as for example, the structure of a molecule containing carbon, hydrogen and oxygen atoms (e.g. α-naphthol), the two aromatic rings have to be drawn, fused and a hydroxyl group fixed in the first position in the ring. This would just be a chemical drawing and could be easily done using software like CHEMWIND,

OH
α-Naphthol

CHEMDRAW, etc. But when the concern is about building a model, we need to breathe in the property of a bond length of 1.33 Å [1 Angstrom = 10^{-10} m] to carbon — carbon bonds in the benzene ring (which are partial double bonds) and of 1.40 Å for a carbon — oxygen bond. These values are obtained through averaging such bonds from X-ray crystallographic studies.[5]

WHAT IS MOLECULAR MODELLING?

Molecular modelling is the scientific art of simulating/mimicking, chemical or biological systems, so that computational methods (*in silico*) can be applied to understand the process concerned. Understanding such processes, does lead to speeding up or slowing it down, on practical lines, as necessity demands. For example, if the molecular process involved in the inflammation of a part in the human system can be understood at the biological level, a methodology can be evolved to moderate this process. A further assumption is that these differences can be expressed in quantitative terms. A modelling study may then aim to find an empirical equation that would relate the structures of compounds to their biological properties, for example the strength of a molecular interaction.

For example, in certain types of inflammation, which on setting in, trigger the production of an enzyme cyclooxygenase (COX-I/COX-II). This enzyme interacts with an endogenous neurotransmitter and triggers the signalling process to the brain to feel the pain.

Models using computers are generated using mathematical equations (classical and quantum mechanical) and are evolved based on experimental information that has to be taken into consideration during model building. The basic assumption in mathematical modelling of molecular interactions is that the biological properties of any molecule — from small organic compounds to bio-macromolecules —

are completely determined by the chemical structure of that molecule. Differences in biological properties ultimately arise from structural differences. Scientists know that the critical feature of a protein is its ability to adopt the right shape for carrying out a particular function. But sometimes a protein twists into the wrong shape or has a missing part, preventing it from doing its job. Many diseases, such as **Alzheimer's** and **Madcow**, are now known to result from proteins that have adopted an incorrect structure.[6]

A molecular model of a small molecule or protein is built on already known parameters that are known to capture the ideal three-dimensional structure of the molecule. Thus, studied (and averaged) information is used as the stepping stone in model building and each step needs to be validated.

Computer models mathematically represent,

1. *The atomic positions* which denote the coordinates and molecular geometry.

2. *Molecular surfaces* which graphically represent the molecule whose surface can be explored.

3. *Energies of a molecule or a system* which is vital for evaluating its stability to exist in that state.

Molecular modelling covers the areas of automatic structure generation, analysis of three-dimensional databases (**structural bioinformatics**), predicting the three-dimensional structure of proteins from their sequence (**homology modelling**), etc. Computational calculation of energies of a system, energy minimization, molecular dynamics and Monte Carlo simulations too fall under this category.

Representations in molecular modelling

The structural representation of molecules include **Connolly Surface**, **Space filling models** (Corey–Pauling–Koltmun (CPK models), **Ball and stick lines**, etc. which are represented below. These models can be coloured according to the nature of their type, electron density, etc.

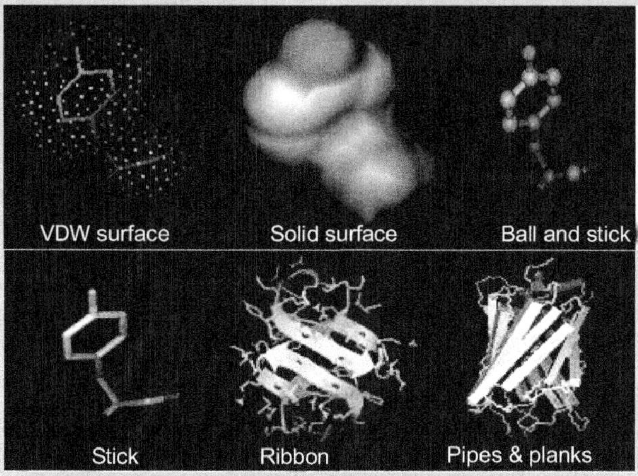

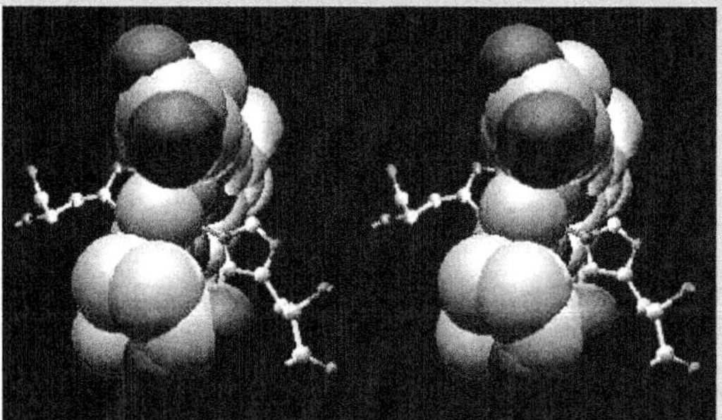

(*See* Plate 1)

Another representation is the stereo-view (given above[7]) for which stereo-glasses can be used. Visualizing two superimposed molecules (having similar skeleton with variation in the functional groups) will indicate the changes in conformation. (View the right diagram with the right eye and the left, with the left eye. Slowly cross-eye (squint view) so that both the images merge together, which on viewing would show the atoms that go below the plane and above the plane defining the stereochemistry).

DRUG DISCOVERY: THE EVOLUTION AND PROCESS

Drugs form an integral part of the human life cycle in the present-day world which is targeted by diseases and epidemics. A drug may be defined as "a chemical entity that, when consumed/injected, results in the control or eradication of a particular disease/infection." A drug will exert its activity through interactions at one or more molecular targets. When the drug does something other than what it is supposed to do, this results in undesirable side effects.

Drug discovery is a pipeline process involved in the evolution of drugs and involves "Genes to Drugs" strategy (Figure 1.1). Identifying the gene responsible for a particular disease state, confirming that a particular protein(s) is/are involved in the disease process and finally evolving a drug to combat the disease—these three form the main areas in this strategy.

Brief History of Drug Discovery

The evolution of drug discovery started in 1874 with Paul Ehrlich at the University of Strasbourg postulating the existence of "chemoreceptors."

1. **Papaverin** (from *Papaver somniferum*) was isolated in 1848, and its antispasmodic properties were discovered in 1917.[8]

2. **Penicillin** was discovered in 1929 by Alexander Fleming,[9] and a large number of antibiotic substances had been described in the scientific literature between 1877 and 1939.

3. The description and characterization of carboanhydrase in 1933[10] was fortuitously followed by the discovery that sulphanilamide, the active metabolite of the sulphonamide (sulpha drug) Prontosil, inhibited this enzyme and that this effect led to an increase in natriuresis and the excretion of water.[11]

4. **Ivermectin,** a superior anti-parasite medication; **Lovastatin,**[12] a hypolipidaemic agent isolated from *Aspergillus terreus,* the immuno-suppressants Cyclosporin A[13] and FK 506[14] were all discovered in the following years.

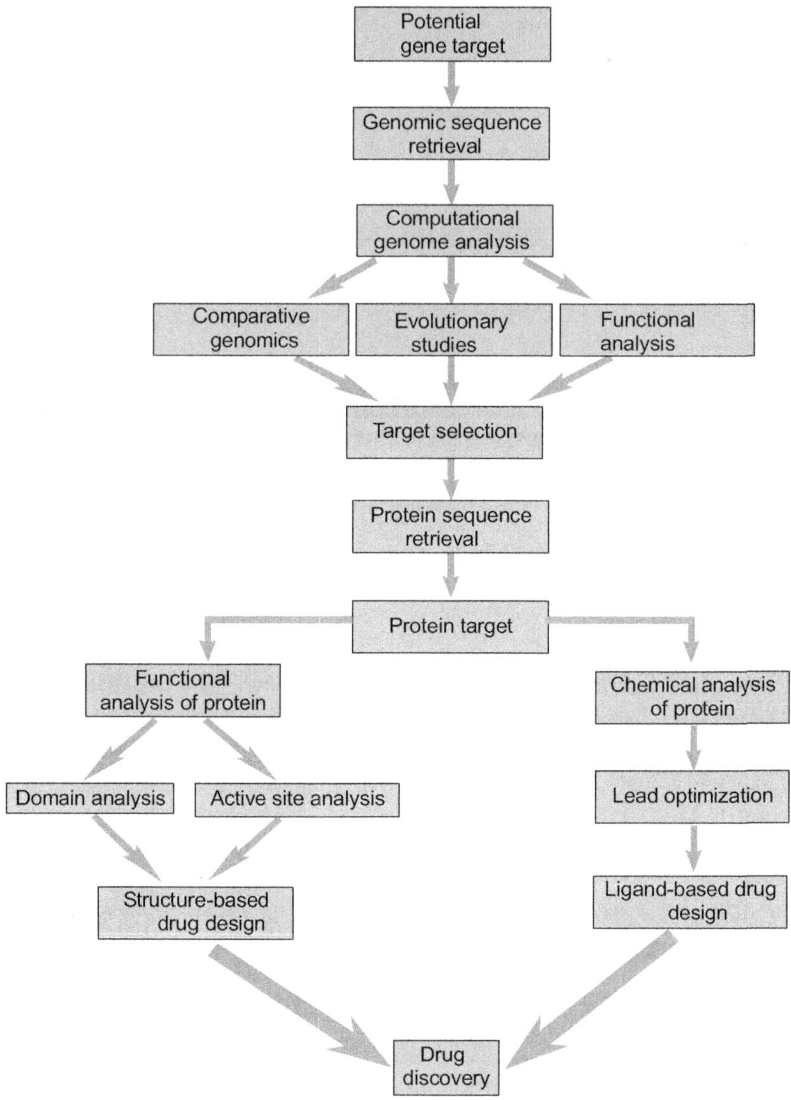

Figure 1.1   "Genes to drugs" strategy

Reducing the research timeline in the discovery stage is a key priority for pharmaceutical companies worldwide. Many companies are trying to achieve this goal through the application and integration of advanced technologies such as computational biology, chemistry, computer graphics and higher performance computing (HPC). Molecular modelling has emerged as a popular methodology for drug design in the sense it can combine computational chemistry and computer graphics. It aims at computer-aided techniques for the efficient identification and optimization of novel molecules with a desired biological activity.

The use of computers in the discovery of effective biological molecules in treating diseases is termed as Computer Assisted Drug Design (CADD). Alternatively, Computer Assisted Drug Design (CADD) is anything that requires the use of computers to paint, describe or evaluate any aspect of the structure of a molecule. Traditionally, the approach towards evolving a drug has been to synthesize molecules and assay them for biological activity in wetlab. This is an extended process that can take as many as 15 years from the synthesis of first compound in the laboratory until the therapeutic agent or drug is brought to market (Figure 1.2). According to the Tufts Center for the Study of Drug Development, a new prescription drug costs on an average, $802 million and takes up to 15 years to develop and get the approval of the Food and Drug Administration Agency.

Much of this cost is an amortization of all the other drugs that fail during development. Also there are serious hurdles regarding ease and cost of synthesis, patentability, safety and social need for the new compound.

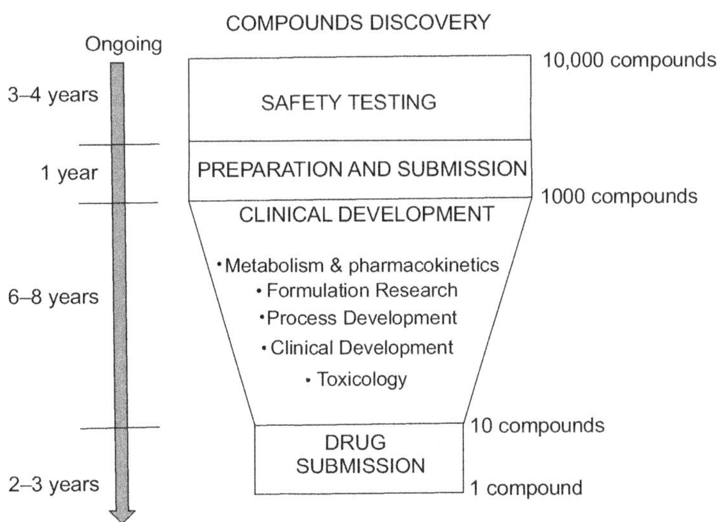

Figure 1.2 Time frame of a drug discovery process

Given the three-dimensional structure of a target receptor molecule, usually a protein, chemical compounds having potential affinity towards it are designed rationally, with the aid of computational methods.[15-20] Thus the major goal in drug research is to find ligands that influence the organism, unwanted cells, viruses or alien bacteria in a way that will lead to desired effects. The inhibition of enzymes or of receptors is a successful approach to influence the activity of the protein. By hindering the naturally binding compounds to dock to the protein, one can stop the metabolic cascade in which the protein is involved. For such purposes, drugs with a very high affinity, i.e., the binding energy, to a protein are necessary. These drugs should bind stronger to the protein than the natural compound that usually interacts with it.

Effective computational approaches may ease down the heavy burden of experimental work. The goals of computation include identification of effective modifications of existing lead compounds as well as discovery of novel lead compounds (de novo design). Although the search for new drugs require intellectual and technological contributions from many scientific disciplines (e.g. chemists, biologists, pharmacokinetists, engineers, etc.), it remains a highly empirical process. The low success rate is due to the imperfect knowledge of biological processes, to our ever-increasing medical objectives (e.g. drugs for oral contraceptives, psychiatric disorders, smoking cessation, etc.), and to the fact that new drugs have to be better than existing drugs.

LEADS, HITS AND DRUGS

Hits are chemical compounds that produce biological activity thought to represent therapeutic potential. They possess the optimum bioactivity, but with some 'to-be improved'/undesirable properties like deactivation (by enzyme) before it reaches the target, poor solubility, etc. Biological screening is carried out to identify those compounds that possess the biological activity (e.g. good receptor binding), better than the '**Hits**'. Such compounds identified are called '**Leads**'. The initial leads are unlikely to be the final drugs. Complex evaluations are necessary, and typically the initial hit is modified atom-by-atom to improve important characteristics of the molecule. The choice of lead structure is very important for success in drug development.

Lead optimization is the process of finding a compound that has some advantage over a related lead. This process can result in a better understanding of the physical–chemical determinants of the newly discovered activity, the reduction of undesirable side effects, experimental verification of the positional requirements of drug–receptor binding, modification of an absorption or metabolic rate, or an increase in the binding coefficient. Underlying the process is the common statistical purpose of characterizing and optimizing a response function. Suitable formulations of optimized compounds are then developed and tested in clinical trials. Because there is no guarantee that a potent compound will become a marketable drug, a large number of new leads are needed to feed into the drug development process. It is desirable to find lead compounds in structurally diverse chemical classes. If multiple chemical classes can be found, they provide optional starting points for lead optimization of activity, physical properties, tissue distribution, plasma half-life, toxicity, etc.

THE ROLE OF COMPUTER-ASSISTED DRUG DESIGN

The target of Computer Assisted Drug Design (CADD) is not to find the ideal drug but to identify and optimize lead compounds and save some experiments. The parameters expected from a drug are

1. Safety

2. Efficiency

3. Stability (chemical and metabolic)

4. Solubility (absorption, distribution)

5. Synthetic viability (bio- or chemo-)

6. Novelty (i.e., patentability)

Properties 2–5 can be evaluated using CADD.

CADD cannot, however, maximize its utility in isolation and will not do so. Rather, it can form a valuable partnership with experiment by providing estimates when experiments are difficult, expensive, or impossible, and by coordinating the experimental data available.

A close coupling between computational chemists and experimentalists allows information to flow immediately and directly between the two. It also provides valuable information for the experimental scientists, guiding to further experimental planning and potentially making this process more efficient.

Drug design using computational methods is dependent on experiments—either experimental result(s) is utilized to arrive at a generalization which can be applied to untested case(s) or a hypothetical model is evolved which is to be experimentally validated.

WHERE DO LEADS COME FROM?

Leads are normally obtained through

- ◈ The study of natural products (e.g. Taxol)

- ◈ Modification of natural products (e.g. Fredericamycin A)

- ◈ Serendipity (e.g. Pencillin)

- ◈ Structure-based drug design (e.g. anti-HIV drugs)

- ◈ Ligand based drug design (e.g. cephalosporin)

- ◈ Chemists intuition (e.g. Enalapril)

Through the early 1940s, much of the process of drug discovery was dependent on plant sources and serendipity.[21] An analysis of the origin of the drugs developed between 1981 and 2002 showed that natural products or natural-product-derived drugs comprised 28% of all New Chemical Entities (NCEs) launched onto the market.[22] Since secondary metabolites from natural sources have been laborated within living systems, they are often perceived as showing more "drug-likeness and biological friendliness than totally synthetic molecules,"[23] making them good candidates for further drug development.[12,13] Current drug discovery from plants has mainly relied on bioactivity-guided isolation methods, which, for example, have led to discovery of the important anticancer agents, paclitaxel from *Taxus brevifolia* and camptothecin from *Camptotheca acuminata*,[24–26] etc. (Figure 1.3).

Taxol from *Taxus* spp. (Yew trees) as a treatment for ovarian cancer is now in the top 30 best-selling drugs.

The technology base in drug discovery is synthetic chemistry. Semi-synthetic modification of naturally occurring compounds has been known to enhance their biological activity.

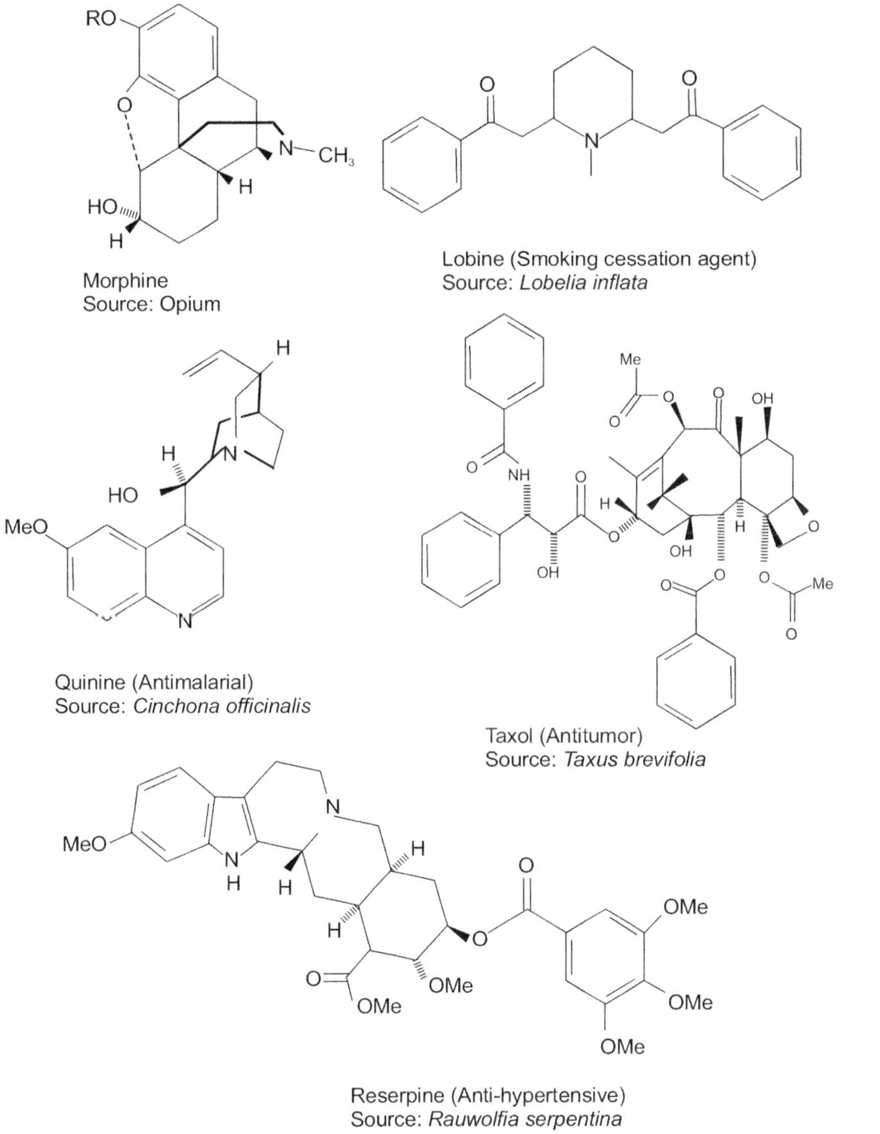

Morphine
Source: Opium

Lobine (Smoking cessation agent)
Source: *Lobelia inflata*

Quinine (Antimalarial)
Source: *Cinchona officinalis*

Taxol (Antitumor)
Source: *Taxus brevifolia*

Reserpine (Anti-hypertensive)
Source: *Rauwolfia serpentina*

Figure 1.3   Chemical representation of some naturally occurring compounds used as drugs

Some compounds show biological activity against more than one type of disease and often times these are discovered during observations during clinical trials.[27] For instance, bupropion hydrochloride better known as Wellbutrin was originally developed as an antidepressant but was found to help patients stop smoking during the clinical trials.

In the early 1990s,[28] researchers at Pfizer were searching for a drug that would reduce the chest pain due to angina. Since an increase in blood and oxygen supply should result in fewer anginas, they looked for a drug that would dilate coronary arteries. They found a substance in the blood, cyclic guanosine monophosphate (cGMP) that caused arteries to dilate and identified an enzyme, phosphodiesterase (PDES), that broke it down. With PDES around, cGMP was destroyed and arteries shrank thin and became blood-poor. Without PDES the arteries opened up. They discovered a molecule that PDES could securely lock onto, thus stopping binding with cGMP. It took approximately four years to find a molecule that allowed PDES to lock on but that the enzyme could not destroy. The drug is now known as **Viagra** and is prescribed for impotence.

A second phase of drug discovery emerged with advances in enzymology and protein biochemistry. Many of the biological pathways and processes were identified as enzymologists and pharmacologists defined new enzymes, receptor ligands, and their functions. As a result, although serendipity was still a major factor, drugs were now directed toward distinct molecular targets.

The third phase of drug discovery is one represented by the popular vision of computer-driven discovery and development of compounds that not only treat the disease, but also have minimal side effects.

HOW DO DRUGS ACT?

The mode of action of a drug is a complex process and usually deciphered using systems biology. When a drug is administered orally or intravenously, the drug has to undergo **absorption** and **dissolution** to reach the area to be acted upon. After the action or during the process, it would be **metabolized** and finally **excreted** from the system. The drug should also possess negligible **toxicity**. All these are termed as the ADMET property of a drug and are the foremost to be evaluated for a drug candidate's success. This is studied through **pharmacokinetics**. Absorption of the drug should finally end in excretion; the in-between process can be monitored through, for example, high-performance liquid chromatography (HPLC).

These most important processes in pharmacokinetics are abbreviated by the acronym LADMET.

- ◈ **Liberation (L)**   If the drug is in a solid phase, it first has to get dissolved.

- ◈ **Absorption (A)**   To enter the bloodstream the drug has to cross different barriers by passive diffusion. These barriers are usually membranes in the duodenum. Highly polar and large molecules have difficulties crossing the lipid bilayer membranes.

- ◈ **Distribution (D)**   The bloodstream will not distribute the medicament equally in the human body. Because of different barriers in the human organism (for example the blood-brain barrier), a drug may not reach some parts of the human organism. Consequently, the drug concentration will be higher in some regions (organs) than in others. Additionally, the distribution of the molecule often also depends on how good a drug flows in the blood. Molecules that are too large will have difficulties to flow quickly to the target and the chances of degeneration are higher.

- ◈ **Metabolism (M)**   There are special enzymes in the human body that detect alien substances and convert them into products which are easy to excrete.

- ◈ **Excretion (E)**   It describes the process of how the drug leaves the human body.

- ◈ **Toxicity (T)**   This would be a major factor in deciding the successful emergence of a drug.

In all living systems, the interaction of a molecule (ligand, e.g. neurotransmitters) with a receptor (proteins) is the source of many biological activities as well as diseases. After passing through the membrane, the ligand/drug docks into the cavity (known as the binding site) in the receptor. Understanding such interactions will give insight into the mechanism of action, which can pave way for methods for controlling such processes. A drug is discovered after undergoing many filters to have its desirable action (Figure 1.4).

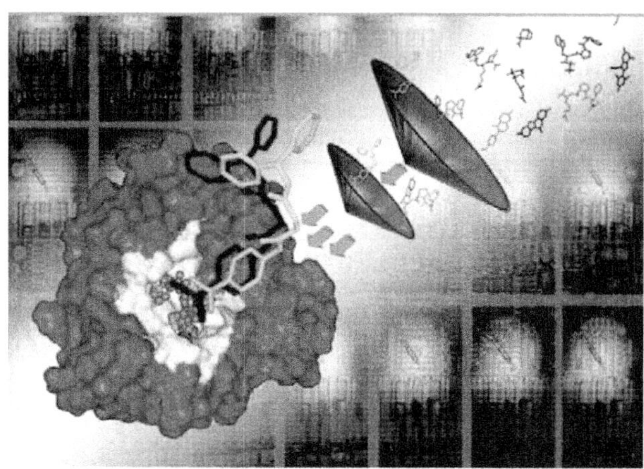

Figure 1.4   A schematic view of the filters that a compound has to pass through to reach the target

Targets

A drug can be an agonist (produces an action) or an antagonist (opposes an action) based on its effect. The target for a drug is the macromolecular species that control the functions of cells. It may be surface-bound proteins like receptors and ion channels or species internal to cells, such as enzymes or nucleic acids (Figure 1.5). DNA can also function as a target, as in cases like netropsin binding preferentially to sequences rich in A–T pairs.

Even the most modest estimates of the total number of proteins encoded by the human genome suggest around 30,000 proteins (not accounting for multiple gene product variants arising from alternative gene splicing and single nucleotide polymorphism). Such a multitude of entities emphasizes the importance of selectivity for molecular interactions. So-called "drug-like" chemical compounds are estimated to be able to access and influence over 3,000 protein–ligand binding sites in humans.[29–30]

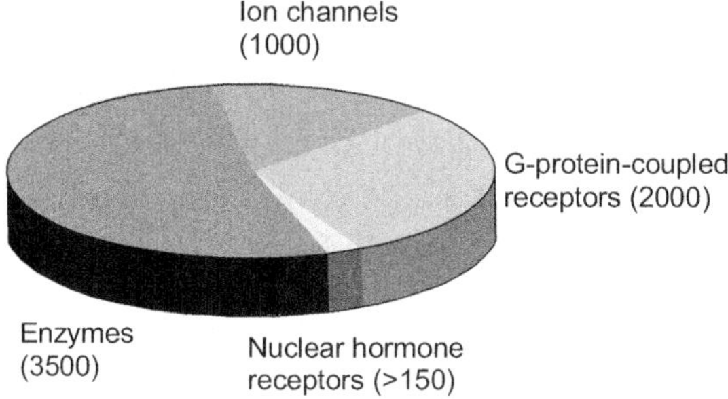

Figure 1.5   Statistics of reported targets (*See* Plate 2)

Proteins are made up of amino acids and the sequence in which a protein is made of is termed as the two-dimensional structure of the protein. Based on the composition and the nature of the amino acids (hydrophobic, polar, etc.), the protein folds to what is known as the **three-dimensional structure**. The structure and function of a protein are interrelated, thus understanding the three-dimensional structure of a protein and looking for a similar structure in a database constitute the **structural bioinformatics** research. If the structures of two proteins isolated from different species are identical, then it is possible that their function will be similar. The similarity in identity is measured by superimposing the two structures and checking in the root mean square deviation (RMSD) value. Lesser the value (say 0.00001Å) closer they are in being identical.

The three-dimensional structure of the protein is what is important for a drug design process. The drugs, while interacting with a protein, dock onto a cavity in the protein called the **binding site** and due to some intermolecular forces, bring in a change in the conformation of the binding site, which might trigger or suppress a cascade of cellular processes. A majority of drugs act by binding with high affinity and great specificity to a particular site on a macromolecular target. Although the binding of such ligands may induce a change in the conformation of the receptor, knowledge of the shape and chemical nature of the binding site should, in principle, allow new molecules with desired properties to be designed.

Under certain conditions, a protein may have a stable alternate conformation, or shape, that enables it to carry out a different biological function. Proteins that exhibit this characteristic are called **allosteric**. The interaction of an allosteric protein with a specific cofactor, or with another protein, may influence the transition of the protein between shapes. In addition, any change in conformation brought about by an interaction at one site may lead to an alteration in the structure, and thus function, at another site. This type of transition affects only the protein's shape, not the primary amino acid sequence. Allosteric proteins play an important role in both metabolic and genetic regulation. For example, in the case of nicotinic acetylcholine receptor, the shape of the protein is altered when **Ivermectin**, a drug, binds to it, and this facilitates the binding of a second molecule.

THE PROCESS OF DRUG DISCOVERY

The traditional protocol for the drug discovery process is to synthesize molecules with variation(s) in their substituents and check for their bioactivity in the wetlab. This would indicate the most potent compound that would then be synthesized on a large scale. In this context the macromolecule, which the drug interacts with is ignored. However, with the evolution of computers, visualization and computation, the process has gained strength. It is now possible to simulate such processes and generate "Hits" just by virtual screening.

File Formats and Molecular Visualization

The foremost aspect of molecular modelling and drug design is the feasibility to view the molecules—protein or ligands. The (molecular) file format describes the layout of a computer data file. It is a set of instructions on how a molecule is encoded with respect to its connectivity, atom types and coordinates, and may contain bibliographic data.

SMILES (Simplified Molecular Line Input Entry Simplification) notation is one of the commonly encountered format in modelling. Each atom is labelled by its symbol and the molecule represented as a string (Table 1.1). This depicts only the connectivity and no information on three-dimensional arrangement of atoms is represented.

Table 1.1   Molecules and their SMILES notation

Molecule	Structure	SMILES
n-Butane	$CH_3-CH_2-CH_2-CH_3$	CCCC
Isobutane	$CH_3-CH-CH_3$ (with CH_3 branch)	CC(C)C
Naphthalene		c1ccc2ccccc2c1
1,3-cyclohexadiene		C1=CC=CCC1

The three-dimensional structure of a molecule can be defined in terms of their internal coordinates, which relates the position of the atoms with respect to each other. The commonly used three-dimensional formats include **pdb** (Table 1.2), **mol2** (Table 1.3), **sdf**, **xyz**. The cif format is the Crystallographic Information File and contains crystallographic information such as space group, wavelength of radiation used, crystal dimensions, etc., apart from the coordinates.

Table 1.2   pdb format for benzene

REMARK 4							
ATOM	1	C1	BEN	1	−1.853	14.311	16.658
ATOM	2	C2	BEN	1	−2.107	15.653	16.758
ATOM	3	C3	BEN	1	−1.774	16.341	17.932
ATOM	4	C4	BEN	1	−1.175	15.662	19.005
ATOM	5	C5	BEN	1	−0.914	14.295	18.885
ATOM	6	C6	BEN	1	−1.257	13.634	17.708
ATOM	7	C7	BEN	1	−2.193	13.627	15.496
END							

Table 1.3   mol2 format for a few atoms in histidine

1	N1	−1.0947	0.5371	1.7186	N.4 1	<1>	0.2252
2	C2	−0.9885	0.9170	0.2765	C.3 1	<1>	0.0213
3	C3	−0.2043	−0.1565	−0.4766	C.3 1	<1>	0.0354
4	C4	−2.3725	1.0376	−0.3154	C.2 1	<1>	0.0897
5	O5	−2.7546	2.1336	−0.8057	O.co2 1	<1>	−0.5442

The mol2 format includes the charge on each atom apart from the coordinates.

Open label is free tool for the interconversion of these formats.

Some Free/Open Visualization Tools

Table 1.4, lists some of the free visualization tools available in the internet along with the addresses.

Table 1.4   Free downloadable visualization tools

Name	Web address for download	Comments
CHIME	http://www.umass.edu/microbio/chime/getchime.htm	A plug-in tool
Rasmol	http://www.umass.edu/microbio/rasmol/getras.htm	A downloadable tool
Spdbv	http://www.expasy.ch/spdbv/	**DeepView** Swiss-PdbViewer is an application that provides a user-friendly interface allowing to analyse several proteins at the same time.
VMD	http://www.ks.uiuc.edu/Research/vmd/	Visualizing Molecular Dynamics
Accelrys DS visualizer	http://www.accelrys.com/products/downloads/ds_visualizer	Can depict intra-molecular hydrogen bonds apart from other functions

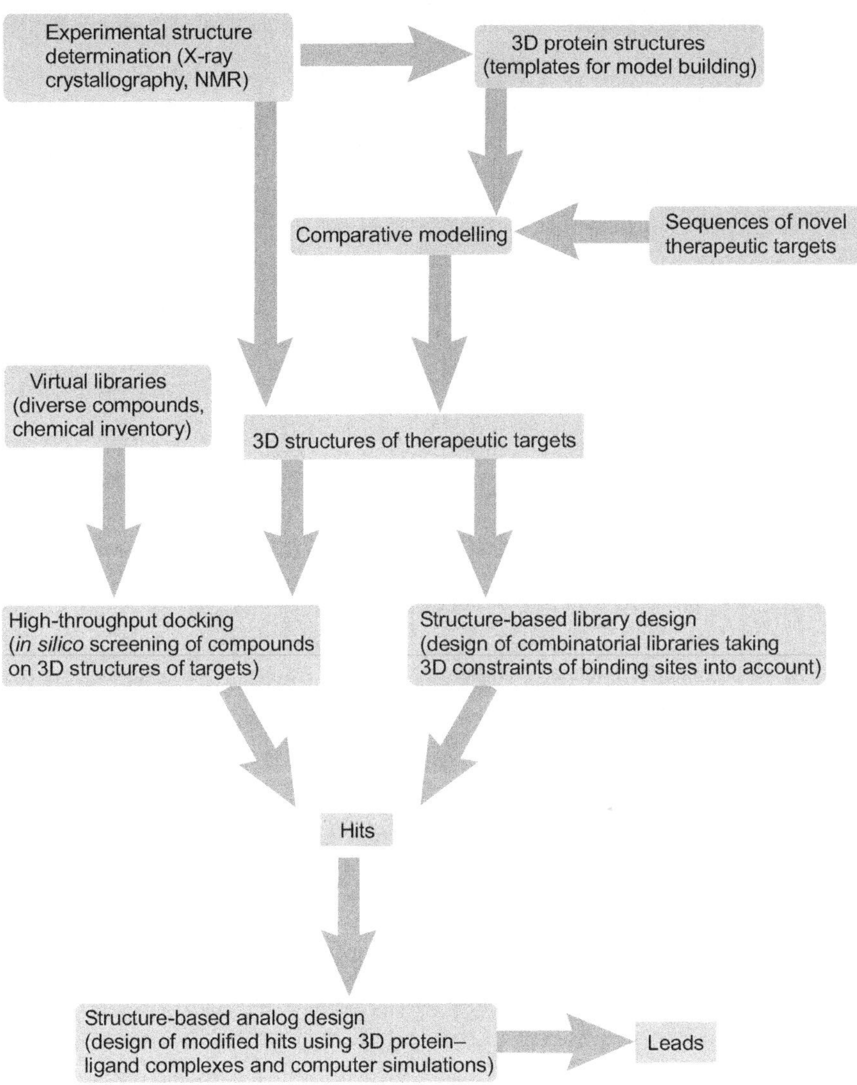

Figure 1.6 The process of drug design

In the traditional drug discovery process, a library of compounds (say 20–30 compounds) are synthesized and their activity against a particular disease is tested using an animal model or wetlab test. For example, the anti-inflammatory potency of a drug is evaluated by the 'paw oedema' test. The paw of the mice is smeared with an agent which induces an inflammation. Then after the gestation period, the drug which is to be evaluated is applied (at specified concentrations). The percentage of efficacy in combating the inflammations is evaluated, along with a standard drug.

The starting point for a drug discovery process is an identified target (typically a protein) and a compound known to interact with it in controlling the process/disease. The overall process (Figure 1.6) includes

1. Target identification

2. Target validation

3. Lead discovery—Identify a ligand that would inhibit the disease process

4. Lead optimization

5. Assays

The Methodology

Historically, the discovery of novel drugs has been led by chemistry and pharmacology. With the advent of genomic sciences, however, biology has established itself as the main driver. Drug-discovery programme typically start with the identification of suitable drug targets. Such targets are bio-molecules, which, in most cases, are proteins, such as receptors, enzymes and ion channels. During the stepwise process of **target validation**, a sufficient level of 'confidence' has to be established that the target is of relevance to the disease under study and modulation of the target will lead to effective disease treatment. The initial steps of target validation are usually obtained *in vitro* and in animal models but the ultimate validation can be achieved only in clinical experiments in humans. After initial target validation has been obtained, modulators of the target have to be identified. Such modulators can be agonists or antagonists in the case of receptors, activators or inhibitors of enzymes, and openers or blockers of ion channels. This phase called **lead-identification** starts with the design and development of a suitable assay to monitor the target under study. Subsequently, high-throughput screening (HTS) exposes the target to a large number of chemical compounds (typically in the order of 10^5) that increasingly come from high-speed parallel and combinatorial synthesis.

Active compounds that demonstrate dose-dependent target modulation are called lead compounds when a certain degree of selectivity for the target under study can be shown and the first positive results in animal models are obtained. Such lead compounds are optimized (**lead optimization**) in terms of potency and selectivity as well as physico-chemical properties, and their pharmacokinetic and safety features are assessed before they can become candidates for drug development.

Given the three-dimensional structure of a target receptor molecule, chemical compounds having potentially high affinity for it are designed rationally, with the aid of computational methods. Effective computational approaches may release the heavy burdens on experimental work. The goals of computation include identification of effective modifications of existing lead compounds, as well as discovery of novel lead compounds (de novo design).

Strategies for CADD vary depending on the extent of structural and other information available regarding the target (enzyme/receptor) and the ligands. "Direct" and "indirect" design are the two major modelling strategies currently used in the drug design process. In the indirect approach the design is based on comparative analysis of the structural features of known active and inactive compounds. In the direct design, the three-dimensional features of the target (enzyme/receptor) are directly considered (Figure 1.7).

		Target (enzyme/receptor)	
		Unknown	Known
Compounds (inhibitor/ligand)	Unknown	Similarity searching Rational screening	De novo design
	Known	Analog-based drug design	Structure-based drug design

Figure 1.7 Four major cases in CADD also known as "direct" and "indirect design when the structure of the target is respectively known or unknown.

ELUCIDATING THE THREE-DIMENSIONAL STRUCTURE

Two main techniques are used for bio-molecular structure determination: X-ray crystallography and nuclear magnetic resonance (NMR) spectroscopy. All resulting structure models are derived from their underlying experimental data.

NUCLEAR MAGNETIC RESONANCE SPECTROSCOPY

Nuclear magnetic resonance (NMR) spectroscopy is unique among the methods available for three-dimensional structure determination of proteins and nucleic acids at atomic resolution, since the NMR data can be recorded in solution.

Considering that body fluids such as blood, stomach liquid and saliva are protein solutions where the protein molecules perform their physiological functions, knowledge of the molecular structures in solution is highly relevant. In the NMR experiments, solution conditions such as temperature, pH and salt concentration can be so adjusted as to closely mimic a given physiological fluid. Conversely, the solutions may also be changed to quite extreme non-physiological conditions, for example, for studies of protein denaturation. Furthermore, in addition to protein structure determination, NMR applications include investigations of dynamic features of the molecular structures, as well as studies of structural, thermodynamic and kinetic aspects of interactions between proteins and other solution components, which may either be other macromolecules or low-molecular-weight ligands.[31] It has thus found widespread use in the field of molecular recognition and supra-molecular chemistry,[32] a natural consequence of this being its utilization in structure-based drug discovery.[33]

The NMR parameters which one can obtain from a single NMR experiment (chemical shifts, coupling constants, signal intensities and line widths) are intimately dependent on the very precise chemical environment of each nucleus of the molecule. With careful experimental control (temperature, concentration, solvent, pH and ionic strength if applicable), the modification of the NMR spectrum by addition of a second compound to the first sample clearly indicates the formation of a complex.

The most commonly employed method to detect interaction between a drug candidate and a target is the chemical shift perturbation method,[34,35] which, as its names suggests, analyses the chemical shift changes observed in the target when a ligand interacts with its surface. To predict the change in the conformation of the protein, due to its interaction with the ligand, it requires the assignment of chemical shifts to the native (non-complexed) protein. The NMR experiments more commonly used for this purpose are the 2D [^{1}H–^{15}N] or [^{1}H–^{13}C]-HSQC/HMQC (Heteronuclear Single Quantum Correlation/ Heteronuclear Multiple Quantum Correlation) in the absence and presence of ligand.

While the former allows detection of changes in the amide protons and nitrogen nuclei of the backbone and Asn and Gln side chains and requires the protein sample to be enriched in ^{15}N, the latter requires ^{13}C enrichment but yields information on chemical shift changes in all side chains. Although ^{13}C labelling allows the chemical shift perturbation approach to sample hydrophobic patches on the surface of the protein, the ^{15}N experiment is normally preferred because it requires neither the relatively costly ^{13}C enrichment nor the often lengthy process of side chain assignment. In both cases, by measuring the chemical shift changes as a function of ligand concentration, the affinity constant between the ligand and the target can be accurately measured.

EXPERIMENTAL DESIGN

The protein may be dissolved in 0.5 ml of water and the ionic strength, pH, and temperature may be adjusted so as to ensure near-physiological conditions (it is advantageous to work in the slightly acid pH range from 3 to 5). The protein concentration should be at least 1 mM, ideally 3–6 mM, so that 15–30 mg of a protein with molecular weight 10,000 should be available for a structure determination. Although this concentration is high relative to that of most proteins in their physiological milieu, it is not far from the total protein concentration in many body fluids. So far, structure determinations by NMR have been reported for proteins with molecular weights up to approximately to 40,000 kDa.

X-RAY CRYSTALLOGRAPHY

X-ray crystallography is a very powerful technique used in determining the three-dimensional structure of a molecule, be it a small molecule or a protein.

X-rays whose wavelength is at the Angstrom level ($1\text{Å} = 10^{-10}\,\text{m}$) are the ideal probes to measure the interaction between atoms (both intra and inter), whose molecular distances too range in the same scale. X-rays are passed through the crystal, and each atom in the protein causes the rays to diffract differently. This diffraction pattern is then recorded by a sensitive detector. About 10 to 100,000 measurements are made in different areas of the crystal, allowing the researcher to gain a complete picture of its structure. After a series of complex calculations, the result is a beautiful photograph of the ligand/protein in three dimensions, which can be displayed and manipulated on a computer.

In the output of such a study, we obtain the coordinates of the atoms present in the molecule which when joined by appropriate bonds, would give rise to the three-dimensional chemical structure of the molecules in crystal. The steps involved in X-ray crystal structure study uses a single crystal for its diffraction study. Powder diffraction is another method for studying the amorphous nature of a metal, etc.

The X-ray diffraction study protocol consists of

1. Crystallizing the compound/protein
2. Checking the crystalline nature of the molecule and its size
3. Collecting data
4. Processing the data
5. Structure solution
6. Structure refinement

With the evolution of computer, the solving of a crystal structure is only a matter of six hours and with necessary expertise, the structure can be solved within a few hours after that. The only drawback could be that the molecule has to be crystalline in nature (a crystal is usually tested under a polarizing microscope and shows extinction) with an approximate size of 0.30 mm in all the three dimensions. The various crystallization methods include the vapour diffusion method, sitting drop method and slow evaporation method for small molecules and the hanging drop method and batch method for proteins.

X-ray crystallography is the only method for determining the "absolute" configuration of a molecule and is the most comprehensive technique available to determine the structure of any molecule at atomic resolution. Results from crystallographic studies provide unambiguous, accurate, and reliable three-dimensional structural parameters, which are prerequisites for rational drug design and structure-based functional studies.

All methods have strengths and weaknesses, and with regard to structure determination via NMR or crystallography, 3D conformation of molecules may be under the influence of crystal packing effects and experimental conditions (pH, protein concentration, etc.). For instance, if the energy barrier between two conformations is low, different conformations can be found under different experimental conditions. Given that it is now common knowledge that enzymes are capable of adapting to a variety of conformations, it is clear that crystal structures of enzymes are only a "snapshot" of a brief moment in time and do not represent the full mobility of the enzyme. As a result, what crystallizes out may not be the structure that is responsible for the activity.

In a similar vein, often only one conformation is available for structural analysis, and it is difficult to know if this structure represents the true *in vivo* conformation or results from a given set of experimental conditions. Also, some regions are flexible, and side chains can re-orient. Furthermore, rules and programs that apply to soluble proteins are not always appropriate to study membrane proteins.

But determining the 3D structures of proteins constitutes a bottleneck, in particular for integral proteins. The use of NMR is limited to quite small proteins, and does not work at any practical level for integral proteins. It is also generally acknowledged that membrane-embedded proteins are not readily amenable to existing crystallization methods for determining X-ray structures,[36] due to their apolar nature, which makes them less crystallizable. Thus, for example, only one X-ray structure of a G-protein-coupled receptor, namely that for bovine rhodopsin, has been determined to atomic resolution[37] (Palczewski *et al.*, 2000). Although the creation of homology models of related proteins could then be attempted, the accuracy of modelling and the quality of current docking algorithms and scoring functions are most often insufficient to make such attempts very useful.

The advantages of protein crystallography include

- high resolution image of all atoms in the protein

- no size limitations (up to 998 kDa, and even virus particles!)

- very easy to study complexes

- ordered water molecules are visible in the experimental data and are often useful in drug lead design

The major drawback includes the need of a crystallizable protein and since the protein molecules are packed in the crystal, it is arguable whether the conformation would be the same in the solution state.

Thus the three-dimensional structures are elucidated using X-ray Diffraction (XRD), Nuclear Magnetic Resonance Spectroscopy (NMR) and Electron Microscopy. If these techniques cannot be applied due to certain constraints such as non-crystallization of a protein, then **homology modelling** is followed to predict the three-dimensional structure. This is the case with G-protein-Coupled Receptors (GPCRS) which are membrane proteins.

Crystallography and Drug Design

One important aspect concerning all the natural processes is that that the molecules tend to interact in their least energy conformation, i.e., the conformation in which there is less intramolecular repulsion among the atoms. The structure obtained through X-ray diffraction study would be the one that would correspond to one of the least energy conformations. Hence crystal structure is an important starting point for molecular modelling or drug design.

Usually, the binding site in a protein is reliably identified by a process called co-crystallization, wherein the isolated protein and the drug (known through other studies to interact with the protein) are crystallized together. This is done either by soaking crystals of the protein in a solution of the compound, or by growing co-crystals from a solution containing both the protein and the compound. When the co-mixture crystallizes, the drug sits into the cavity (binding site) of the protein molecule. The crystal structure of the protein–ligand complex is then determined by conventional X-ray crystallographic methods.

The 3D atomic model so obtained provides detailed information on the interactions between the protein and the ligand at the atomic level, and so provides an excellent starting point for the synthetic medicinal chemist to begin designing modifications aimed at optimizing both the binding affinity and the compound's pharmacokinetic profile.

AIDS drugs, such as Agenerase and Viracept, were developed using the crystal structure of HIV protease[38,39] and the flu drug Relenza was designed using the crystal structure of neuraminidase.[40] Thus, biomechanistic systems lending themselves to crystallization followed by X-ray diffraction will be readily exploited by structure-based drug design.

Molecular Geometry

The Molecular geometry of a molecule defines its three-dimensional structure in terms of its bond angles, bond lengths and dihedral angle. For example, a $C^{sp3} - C^{sp3}$ hydrogen bond is 1.55Å (inferred from X-ray diffraction studies (Allen *et al.*)). A dihedral angle is defined as the angle formed by the two central atoms B-C in a torsion A-B-C-D.

Conformation and Configuration

Configuration is the arrangement of atoms in space around a particular atom which is the cause of three-dimensional structure. When a carbon is surrounded spatially by four different atoms (or groups) it is said to be an asymmetric carbon which gives rise to the concept of chirality. Pharmacological activity of compounds (drugs) depends mainly on their interaction with biological matrices (drug targets), such as proteins (receptors, enzymes), nucleic acids (DNA and RNA) and biomembranes (phospholipids and glycolipids). All these matrices have complex three-dimensional structures that are capable of recognizing (binding) specifically the ligand (drug) molecule in only one of the many possible arrangements in the three-dimensional space. It is the three-dimensional structure of the drug target that determines which of the potential drug candidate molecules is bound within its cavity and with what affinity.

Chirality is fundamental in biology. Due to the chiral nature of amino acids (except glycine), drug-binding sites of proteins are asymmetric. Thus absolute configuration is critical to proper function in biological systems. Milton and co-workers [41] showed that inverting the chirality of an entire enzyme also inverts its enantioselectivity.

It has been observed that the enantioselectivity exhibited by an enzyme/protein towards a drug is related to the chirality of the drug. For example, the pharmacological properties of ibuprofen reside almost exclusively with the S-(+) enantiomer. More than one-half of marketed drugs are chiral.[42] It is well established that the opposite enantiomer of a chiral drug often differs significantly in its pharmacological,[43] toxicological,[44] pharmacodynamic, and pharmacokinetic [45] properties. Table 1.5 illustrates the importance of chirality in determining the molecular properties:

Table 1.5 A few drugs with enantiomeric selectivity in their biological activity

Thalidomide used as a sedative for pregnant women S-(–) isomer is a strong teratogen, the R-(+) enantiomer is safe and used as an antimicrobial agent in the treatment of leprosy

(Contd.)

Table 1.5 (Continued)

(–) menthol causes the cooling feeling when consumed whereas (+) menthol has no effect

Of the two enantiomeric forms of ibuprofen, only the S-configuration has anti-inflammatory activity

Sensory receptors are G-protein-coupled receptors that differentiate between enantiomers, like all other receptors. Correspondingly, enantiomers can even be recognized by their characteristic odour, e.g. the monoterpenes (R)- and (S)-limonene and (R)- and (S)-carvone,[46] or by their odour intensity. 'Conformation' restricts to ring systems wherein the atoms in the rings can assume various allowed positions with respect to each other. For example, the cyclohexane ring can be in chair or boat conformation, being influenced by the substituents.

(R)-(+)-Limonene	(S)-(–)-Limonene	(S)-(+)-Carvone	(R)-(–)-Carvone
Odour: Orange	Lemon	Caraway	Peppermint

BIOASSAY

Compounds swith desirable characteristics (hits) are identified by bioactivity assays.

Biological assay is a wetlab study in which the biological or the pharmacological activity of compound/drug is evaluated. The test can involve animal models like rat, guinea pigs, etc. and is termed as cell-based assays.

Cell-based assays are generally preferred in drug discovery because the assessment of molecular interactions occurs within the context of a living cellular environment. In addition, information about drug penetration is obtained early on. Cell-based assays can be simple growth-inhibition assays measuring the effect of compounds on cellular growth. Such assays use spectrophotometric or turbidimetric methods for detection of activity.

Biochemical assays have the advantage of providing target-specific information. One of the newer biochemical assays is the capillary electrophoresis (CE) technique that allows detection of functional activity of compounds as well as their relative binding strengths in crude extracts even in the presence of interferences. This approach uses electrophoretically mediated micro-analysis (EMMA), which incorporates laser-induced fluorescence detection for maximum sensitivity. [47] This is a very important aspect in drug design because based on the behaviour of certain chemical compounds towards a particular disease, certain generalities can be made and applied to a set of untested compounds.

LIPINSKI'S RULE OF FIVE[48]

The Lipinski's rule of five has been a critical filter for drug development programs. These principles filter out molecules likely to have poor intestinal permeability or poor aqueous solubility, and hence poor oral absorption. This landmark contribution to drug development has influenced the way that the pharmaceutical industry approaches the development of orally active drugs. Drug discovery programs worldwide use the rules as a filter in high-throughput screening libraries. The Lipinski's rule quantifies the properties that are expected out of a poor drug absorption or permeation which is more likely, when,

1. The molecular weight of the drug is over 500.

2. log P is over 5.

3. There are more than five hydrogen bond donors.

4. There are more than ten hydrogen bond acceptors.

Compounds violating more than one of the Lipinski's Rules are assumed to have problems with bioavailability.

DATABASES

Databases are libraries containing sequences/structural information of proteins and small molecules. Database searching is an important protocol in both molecular modelling and drug discovery. With reference to proteins, structural information of similar homologs is useful in model building while in drug discovery, one of the main utility in database searching is in lead optimization. Predefined parameters such as the presence of certain functional groups are used to obtain reported compounds with the feature.

Databases can be broadly classified into five categories. The **first group** contains resources of structures of low-molecular-weight compounds that are potential ligands. These can be experimentally determined three-dimensional structures of low- molecular-weight-compound, like the CSD[49] (Cambridge structure database) which contains the crystal structures of more than 250,000 molecules.

PUBCHEM is an open database that contains the chemical structure and certain physiochemical properties of chemical compounds that have been reported.

The Available Chemical Database (ACD) (http://www.akosgmbh.de/acd.htm) is a database that has a compilation of chemicals which can be commercially procured. The commercial Asinex database (http://www.asinex.com/) and Open National Cancer Institute (NCI)[50] (http://www.cancer.gov/) database also hold such computed structures. The Chembridge database[51] contains one of the highest number of compounds (700,000) (http://www.chembridge.com/).

The **second group** provides structures of proteins, and here the Protein Data Bank (PDB)[52] is the only player in the field. Research

Collaborators for Structural Bioinformatics is an open databank (www.rcsb.org) which is a repository of the three-dimensional structure of proteins (around 47,400) solved through X-ray crystallography, nuclear magnetic electron microscopy, electron microscopy and such other methods.

Information in the PDB has often been enriched by the so-called "secondary databases" that belong to the **third group.** These databases provide fold classifications,[53] enzymatic functions,[54] links to sequence data.[55] The **fourth group** focuses on databases delivering structures and additional annotation about ligand molecules from the PDB. The chemical and spatial information within ligand structures can be used to refine protein models, specifically to optimize side-chain conformations around binding sites.

HIC-Up[56] comprises chemical and structural information for small molecules—heterocompounds—found in the PDB. Ligand Depot [57] and Relibase [58] provide a graphical interface search among the ligands by two-dimensional similarity and chemical substructure as well as for sequence similarity search among the corresponding well-characterized, known drugs approved by the WHO have been collected in the Super-Drug database.[59] The Chemical Abstracts (CA) provides information on drugs including the CAS-number (useful as cross-reference to other databases) and the chemical 2D structure.

*Ki*Bank[60] is (freely available at http://kibank.iis.u-tokyo.ac.jp/) a database of inhibition constant (K_i) values with 3D structures of target proteins and chemicals. K_i values are accumulated from peer-reviewed literature searched via PubMed. The 3D structure files of target proteins are from Protein Data Bank (PDB). *Ki*Bank has been designed to support structure-based drug design. It provides structure files of proteins and chemicals ready for use in virtual screening through automated docking methods, while the K_i values can be applied for tests of docking/scoring combinations, program parameter settings, and calibration of empirical scoring functions. Additionally, the chemical structures and corresponding K_i values in *Ki*Bank are useful for lead optimization based on quantitative structure–activity relationship (QSAR) techniques.

There are some specialized databases like the SCORPION2[61] database which contains more than 800 records of native and mutant toxin sequences enriched with binding affinity and toxicity information, 624 three-dimensional structures and some 500 references. The internet addresses of some databases are given in Table 1.6.

Table 1.6 Web addresses of some database libraries

No	Web address	Comments
1.	http://redpoll.pharmacy.ualberta.ca/drugbank/	Contains information on a particular drug including the target.
2.	http://www.agklebe.de/affinity	AffinDB is a database of affinity data for structurally resolved protein–ligand complexes from the Protein Data Bank (PDB).
3.	http://apps1.niaid.nih.gov/struct_search/an/an_search.htm	This database contains approximately 114,500 compounds that have been tested against HIV, HIV enzymes, or opportunistic pathogens.
4.	http://aps.unmc.edu/AP/main.php	An antimicrobial peptide database (APD) has been established based on an extensive literature search. It contains detailed information for 525 peptides (498 antibacterial, 155 antifungal, 28 antiviral and 18 antitumor).
5.	http://gdds.pharm.kyoto-u.ac.jp:8081/glida	G-protein-coupled receptors (GPCRs) represent one of the most important families of drug targets in pharmaceutical development.

(Contd.)

Table 1.6 (Continued)

No	Web address	Comments
		GPCR-LIgand DAtabase (GLIDA) is a novel public GPCR-related chemical genomic database that is primarily focused on the correlation of information between GPCRs and their ligands.
6.	http://dtp.nci.nih.gov/docs/ 3d_database/dis3d.html	A searchable database of three-dimensional structures has been developed from the chemistry database of the NCI Drug Information System (DIS), a file of about 450,000 primarily organic compounds which have been tested by NCI for anticancer activity.
7.	http://www.orpha.net/	ORPHANET is a database dedicated to information on rare diseases and orphan drugs.
8.	http://bioinf.charite.de/superdrug/	The Super Drug Database contains 2.396 compounds with 108.198 conformers.
9.	http://bioinformatics.charite. de/supernatural/	A database of 50,000 natural compounds from different suppliers.
10.	http://xin.cz3.nus.edu.sg/ group/cjttd/ttd.asp	A database to provide information about the known and explored therapeutic protein and nucleic acid targets,

(Contd.)

Table 1.6 (Continued)

No	Web address	Comments
		the targeted disease, pathway information and the corresponding drugs/ligands directed at each of these targets. Also included in this database are links to relevant databases that contain information about the function, sequence, 3D structure, ligand binding properties, enzyme nomenclature and related literatures of each target."
11.	http://ligand.info/	Ligand.Info is a compilation of various publicly available databases of small molecules such as ChemBank, ChemPDB, KEGG, NCI, AKos GmbH, Asinex Ltd, and TimTec. The total size of the Meta-Database is 1 million entries. The compound records contain calculated three-dimensional coordinates and sometimes information about biological activity. Some molecules have information about FDA drug approving status or about anti-HIV activity. Meta-Database can be downloaded in SDF format and used for virtual high-throughput screening of new potential drugs.

REFERENCES

1. Schapira, M., Raaka, B.M., Samuels, H, H. and Abagyan, R. (2000). "Rational design of novel nuclear hormone receptor antagonists." *PNAS* 97 (3): 1008–1013.

2. Wang, X. and Richards, N.G.J. (2005). "Computer-based strategy for modeling the interaction of AGRP and related peptide ligands

with the AGRP-binding site of murine melanocortin receptors." *Curr. Pharm. Des.* 11: 345–356.

3. Colmenarejo, G. (2005)."*In silico* ADME prediction: data sets and models." *Curr. Computer-Aided Drug Des.* 1: 365–376.

4. Jennings, A. and Tennant, M. (2005). "Discovery strategies in a biopharmaceutical startup: Maximising your chances of success using computational filters :Computer aided drug design. "*Curr. Pharm. Des.* 11: 335–344.

5. Allen, F.H., Kennard, O., Watson, D.G., Brammer, L., Orpen, A.G. and Taylor, R. (1987). *J.Chem. Soc. Perkin Trans.* 2: S1–19.

6. Tapan, K. Chaudhuri and Subhankar Paul. (2006). "Protein-misfolding diseases and chaperone-based therapeutic approaches." *FEBS Journal* 273 (7): 1331–1349.

7. http://www.usm.maine.edu/~rhodes/Help/StereoView.html

7. Sneader, W.(1985). *Drug Discovery: The Evolution of Modern Medicines.* Wiley, New York.

8. Fleming, A. (1940). *Br. J. Exp. Pathol.* 10: p.226.

9. Taylor, H.R. (1987). *Int. Ophthalmol.* 2: p.83

10. Alberts, A.W. (1988). "Discovery, biochemistry and biology of lovastatin." *Am. J. Cardiol.* Nov 11, 62(15):10J–15J.

11. White, D.J.G. (ed.).(1982). Proceedings of an International Conference on Cyclosporin A. Elsevier, Amsterdam. pp. 5–19.

12. Tanaka, H., Nakahara, K., Hatanaka, H., Inamura, N. and Kuroda, A. (1997). "Discovery and development of a novel immunosuppressant, tacrolimus hydrate." *Yakugaku Zasshi.* 117(8): p. 542.

13. Meldrum, N.U. and Roughton, F.J. (1933)."Properties carbonic anhydrase. Its preparation and properties." *J. Physiol.* 80: 113–142.

14. Schwartz, W. B. N. (1949). "The effect of sulfanilamide on salt and water. Excretion in congestive heart failure." *Engl. J. Med.* 240:173.

15. Jorgensen, W.L. (2004). "The many roles of computation in drug discovery." *Science.* 303: 1813–1818.

16. David Handelsman. Pharmaceutical Industry, heal thyself. www.sas.com/news/feature/his/mar05dev.html

17. Lunney, E. "Computing in drug discovery: The design phase. IEEE computing in science and engineering magazine." http://computer.org/ cise/homepage/2001/05Ind/ 05ind.htm

18. Ooms, F. (2000). "Molecular modeling and computer aided drug design. Examples of their applications in medicinal chemistry." *Current Medicinal Chemistry.* Vol. 7, No. 2. pp. 141–158.

19. Hubbard, R.E. (1997). "Can drugs be designed?" *Curr. Opin. Biotechnol.* 8: 696–700.

20. Jackson, R.C. (1995). "Update on computer-aided drug design." *Curr. Opin. Biotechnol.* 6: 646–651.

21. Walter, H. Lewis and Memory, P. Elvin-Lewis. (1995). "Medicinal plants as sources of new therapeutics." *Annals of the Missouri Botanical Garden.* Vol. 82, No. 1. pp. 16–24.

22. Newman, D.J., Cragg, G.M. and Snader, K.M. (2003). "Natural products as sources of new drugs over the period 1981–2002." *J. Nat. Prod.* 66:1022–1037.

23. Koehn, F.E. and Carter, G.T. (2005). "The evolving role of natural products in drug discovery." *Nat. Rev. Drug. Discov.* 4 : 206–220.

24. Balunas, M.J. and Kinghorn, A.D. (2005). "Drug discovery from medicinal plants." *Life Sci.* 78 : 431–441 .

25. Drahl, C., Cravatt, B.F. and Sorensen, E.J. (2005). "Protein-reactive natural products." *Angew. Chem. Int. Ed. Engl.* 44: 5788–5809.

26. Kinghorn, A.D. (1994). "The discovery of drugs from higher plants." In: Gullo, V.P. (ed.). *The Discovery of Natural Products*

with Therapeutic Potential. Butterworth-Heinemann. Boston, MA. 81–108.

27. Silverman,R.B. (2004). *The Organic Chemistry of Drug Design and Drug Action*, 2nd edn.Elsevier, Amsterdam. p. 617.

28. James, R. Gardner. (1999). "Discovery New Drugs." *How Things Work, Science and Technology. Research/Penn State.* Volume 20, Number 2.

29. Hopkins, A.L. and Groom, C.R. (2002). "The druggable genome." *Nat. Rev. Drug* Disc. 1: 727–730.

30. Lipinski, C. and Hopkins, A. (2004). "Navigating chemical space for biology and medicine." *Nature.* 432: 855–861.

31. Wüthrich, K. (1986). *NMR of Proteins and Nucleic Acids*. Wiley, New York.

32. Pons, M. and Millet, O. (2001). "Dynamic NMR studies of supramolecular complexes." *Prog. Nucl. Magn. Reson. Spectrosc.* 38: 267–324.

33. Peng, J.W. , Lepre, C.A. , Fejzo, J., Abdul-Manan, N. and Moore, J.M. (2001). "Nuclear magnetic resonance-based approaches for lead generation in drug discovery." *Methods Enzymol.* 338: 202–230.

34. Johnson, J.M., Meiering, E. M., Wright, J. E., Pardo, J., Rosowsky, A. and Wagner, G. (1997). "NMR solution structures of the antitumor compound PT523 and NADPH in the ternary complex with human dihydrofolate reductase." *Biochemistry.* 36: 4399–4411.

35. Kallen, J., Spitzfaden, C., Zurini, M.G.M., Wider, G., Widmer, H, Wüthrich, K. and Walkinshaw, M.D. (1991). "Structure of human cyclophilin and its binding site for cyclosporin-A, determined by X-ray crystallography and NMR spectroscopy." *Nature.* 353:276–279.

36. Renfrey, S. and Featherstone, J. (2002). "Structural proteomics." *Nat. Rev. Drug Discov.* 1: 175–6.

37. Palczewski, K., Kumasaka, T., Hori, T., Behnke, C.A., Motoshima, H., Fox, B.A., Le Trong, I., Teller, D.C., Okada, T., Stenkamp, R.E., Yamamoto, M. and Miyano, M. (2000). "Crystal structure of rhodopsin: A G protein-coupled receptor." *Science.* 289: 739–45.

38. Miller, M. *et al.* (1989). "Structure of complex of synthetic HIV-1 protease with a substrate-based inhibitor at 2.3 ANG. resolution." *Science.* 246: 1149–1152.

39. Lapatto, R. *et al.* (1989). " X-ray analysis of HIV-1 proteinase at 2.7 Å resolution confirms structural homology among retroviral enzymes." *Nature.* 342: 299–302.

40. Varghese, J.N. (1999). "Development of neuraminidase inhibitors as anti-influenza virus drugs." *Drug Dev. Res.* 46: 176–196.

41. Milton, R.C., Milton, S.C.F. and Kent, B.B.H. (1992). "Total chemical synthesis of a D-enzyme: the enantiomers of HIV-1 protease show demonstration of reciprocal chiral substrate specificity." *Science.* 256: 1445–1448.

42. Millership, J. and Fitzpatrick, A. (1993). "Commonly used chiral drugs: A survey." *Chirality.* 5: 573–576.

43. Islam, M.M., Mahdi, J.G. and Bowen, I.D. (1997)."Pharmacological importance of stereochemical resolution of enantiomeric drugs." *Drug Safety.* 17: 149–165.

44. Wainer, I. (1993). *Drug Stereochemistry: Analytical Methods and Pharmacology*, 2nd edn. Marcel Dekker, New York.

45. Lipinski, C.A., Lombardo, F., Dominy, B.W. and Feeney, P.J. (2001). "Experimental and computational approaches to estimate solubility and permeability in drug discovery and development settings."*Adv. Drug Del. Rev.* 46: 3–26.

46. Friedman, L. and Miller, J.G. (1971). "Odor incongruity and chirality." *Science.* 172: 1044.

47. Liu, G., Eggler, A.L., Dietz, B.M., Mesecar, A.D., Bolton, J.L., Pezzuto, J.M. and van Breemen, R.B. (2005). *Anal. Chem.* 77: 6407.

48. Christopher, A. Lipinski, Franco Lombardo, Beryl, W. Dominy and Paul, J. Feeney. (1997). "Experimental and compuutional approaches to estimate solubility and permeability in drug discovery and development settings." *Adv. Drug Delivery Rev.* 23: 3–25.

49. Allen, F.H., Davies, J. E., Gallo, J. J., Johnson, O., Kennard, O., Macrae, C.F., Mitchell, E.M. Mitchell, G.F. Smith, J.M. and Watson, D. (1991). "The development of versions 3 and 4 of the Cambridge structural database system. " *J. Chem. Inf. Comput. Sci.* 31 : 187–204.

50. Ihlenfeldt, W.D., Voigt, J. H., Bienfait, B., Oellien, F. and Nicklaus, M.C. (2002). "Enhanced CACTVS browser of the Open NCI Database." *J.Chem. Inf. Comput. Sci.* 42(1) : 46–57.

51. http://www.chembridge.com/chembridge/data.html

52. Berman, H.M., Westbrook, J., Feng, Z., Gilliland, G., Bhat, T.N., Weissig, H., Shindyalov, I.N. and Bourne, P.E. (2000). "The protein data bank." *Nucleic Acids Res.* 28(1): 235–242.

53. Andreeva, A., Howorth, D., Brenner, S.E., Hubbard, T.J.P., Chothia, C. and Murzin, A.G. (2004). "SCOP database in 2004: refinements integrate structure and sequence family data." *Nucleic Acids Res.* (Database issue). 32:D226–D229.

54. Bairoch, A. (2000). "The enzyme database in 2000." *Nucleic Acids Res.* 28: 304–305.

55. Bairoch, A., Apweiler, R., Wu, C.H., Barker, W. C., Boeckmann, B., Ferro, S., Gasteiger, E., Huang, H., Lopez, R., Magrane, M., Martin, M.J., Natale, D.A., O'Donovan, C., Redaschi, N. and Yeh, L.S. L. (2005). "The Universal protein resource (UniProt)." *Nucleic Acids Res.* (Database issue) 33: D154–159.

56. Kleywegt, G. and Jones, T. (Nov 1998). "Databases in protein crystallography." *Acta. Crystallogr. D. Biol.Crystallogr.* 54: 1119–1131.

57. Feng, Z., Chen, L., Maddula, H., Akcan, O., Oughtred, R., Bermannd, H. M., and Westbrook, J. (2004). "Ligand depot: a

data warehouse for ligands bound to macromolecules." *Bioinformatics.* 20(13) : 2153–2155.

58. Hendlich, M., Bergner, A., Günther, J. and Klebe, G.(2003). "Relibase: design and development of a database for comprehensive analysis of protein-ligand interactions." *J.Mol.Biol.* 326(2):607–620.

59. Andrean Goede, Mathias Dunkel , Nina Mester, Cornelius Frommel and Robert Preissner. (2005). "SuperDrug: a conformational drug database." *Bioinformatics.* Vol.21, No.9. pp.1751–1753.

60. Zhang, J.W., Aizawa, M., Amari, S., Iwasawa, Y., Nakano, T. and Nakata, K. (2004). "Development of *Ki*Bank, a database supporting structure-based drug design." *Computational Biology and Chemistry.* 28 (5–6): 401–407.

61. Paul, T.J. Tan , Anitha Veeramani, Kellathur, N. Srinivasan, Shoba Ranganathan and Vladimir Brusic. (2006). "*Toxicon* SCORPION2: A database for structure–function analysis of scorpion toxins." *Toxicon.* Vol. 47, Issue 3, March 1. pp. 356–363.

Recommended Books for Reading

1. *Molecular Modelling: Principles and Applications* Andrew, R. Leach.

   Amazon Publishers.

2. *Guidebook on Molecular Modeling in Drug Design.* Claude N. Cohen. Academic Press.

Free Websites

1. http://www.usm.maine.edu/~rhodes/SPVTut/index.html

2. http://cost.georgiasouthern.edu/chemistry/general/molecule/tutorial/index.htm

3. www.planaria-software.com/

2

MOLECULAR MECHANICS AND MOLECULAR DYNAMICS

INTRODUCTION

Computational techniques in small/macromolecules aim to compute

- ◈ the geometrical arrangements of atoms corresponding to their least energy conformation

- ◈ the relative energies of different conformations

- ◈ the atomic/molecular properties (dipole moment, vibrational spectrum, etc.)

- ◈ time dependence of structure and properties

The two main simulation techniques in explaining the structure and the reactivity are Quantum Mechanics and Molecular Mechanics. While quantum mechanical calculations can effectively describe a system containing only a fewer number of atoms (say hundred), molecular mechanics or the force field methods can do so for thousands of atoms. One of the main applications of molecular mechanics is in **energy minimization.**

Molecular dynamics (MD) deals with the time evolution of a molecular system, resulting in a trajectory and basically is simulation of motion. The output in MD is a trajectory of the conformation differing with time, temperature, environment, etc. The time dependence of the motion of a molecule is called its **trajectory.** Using MD it is possible to predict the conformation of molecule(s) in a particular environment (solvent/vaccuo, etc.) at a particular pH at a specific temperature.

Monte Carlo simulation is also a dynamic sampling technique. It analyses the conformational space in terms of thermodynamic properties like entropy, enthalpy, etc.

Molecular dynamics and Monte Carlo simulations of proteins are used as tools to investigate their structure and dynamics under a wide variety of conditions, ranging from studies of ligand binding[1] and enzyme-reaction mechanisms[2] to problems of denaturation and protein re-folding.[3] Fundamental to such simulations is the representation of the energy of the protein as a function of its atomic coordinates.

QUANTUM MECHANICAL SIMULATIONS

Quantum mechanics describes the energy of a molecule in terms of its interaction between the nuclei and electrons based on the Schrödinger equation

$$\left(\underbrace{\frac{-h^2}{8\pi^2 m}\nabla^2}_{\text{Kinetic energy}} + \overbrace{V(x,y,z)}^{\text{Potential energy}} \right) \underbrace{\Psi(x,y,z)}_{\substack{\text{Wavefunction describing} \\ \text{the system}}} = \overbrace{E\,\Psi(x,y,z)}^{\text{Total energy}}$$

The aim of quantum mechanical methods is to obtain an approximate solution of the Schrödinger equation. The exact solution of this equation is possible only for smaller systems like hydrogen atom, helium atom, particle in a box, etc. Such an approach calculates the properties of the molecule by equations of quantum physics, involving interactions between electron and nuclei. Electron movements are more rapid and, since they rotate independently of the nucleus, it is possible to describe electronic energy separately from the nuclear one. Some approximations based on empirical data are made in calculations by this process, which are not exact and may be executed by two methods, ***ab initio*** and **semi-empirical**. The first one is applied only to small molecules, and although more precise and not needing stored data, requires ample computer memory capacity and time. On the other hand, the semi-empirical method is faster and can be used to minimize energy and optimize molecules with 10 to 120 atoms, although less accurate. Energy is calculated by the Schrödinger equation from stored parameters; MOPAC [**M**olecular **O**rbital **Pac**kage] is the most frequently used semi-empirical method, subdivided into the following: AM1(Austin Model 1), MINDO/3 (**M**odified **I**ntermediate **N**eglect of **D**ifferential **O**verlap), MNDO (**M**odified **N**eglect of **D**ifferential **O**verlap), and PM3.

Various approximations have been invoked for tractability in finding an approximate solution for Schrödinger equation, giving rise to *ab initio* and **semi-empirical techniques**.

Ab Initio Methods

The most rigorous, and consequently the most computationally demanding, is the *ab initio* method. *Ab initio* meaning "from first principles," uses rigorous approximations borne from general physical principles alone and requires no experimental parameters. As such, quantitative application is not limited to specific types of system.[4-6]

Ab initio structure prediction seeks to predict the native conformation of a protein from the amino acid sequence alone. The area is based on the belief that the native folding configuration of most proteins corresponds to the lowest free energy of the sequence. Therefore the biggest challenges with regard to *ab initio* prediction are 1) devising a free energy function that can distinguish native structures from incorrect non-native ones, 2) devising a search method to explore the huge conformational space. *Robetta* is one of the most successful *ab initio* systems in recent years http://robetta.bakerlab.org/). It is built upon accumulated domain knowledge of non-homologous sequences and their solved three-dimensional structures and then applies simulated annealing to create protein tertiary structures. However, the overall prediction accuracy using *ab initio* methods is still very low and a reliable free energy function is still under debate.

Ab initio calculations can be applied to

◈ Organics, organo-metallics, and molecular fragments (e.g. catalytic components of an enzyme)

◈ Vacuum or implicit solvent environment

◈ Study ground, transition, and excited states (certain methods)

Specific implementations include: GAMESS, GAUSSIAN, etc.

One of the main applications of *ab initio* method, is in homology modelling, wherein the three-dimensional structure of a protein is predicted from its amino acid sequence. There are three major approaches to this technique comparative modelling, threading, and *ab initio* prediction.

Comparative modelling exploits the fact that evolutionarily related proteins with similar sequences, as measured by the percentage of identical residues at each position based on an optimal structural superposition, often have similar structures. For example, two sequences that have just 25% sequence identity usually have the same overall fold.

Threading methods compare a target sequence against a library of structural templates, producing a list of scores. The scores are then ranked and the fold with the best score is assumed to be the one adopted by the sequence.

Finally, the *ab initio* prediction methods consist in modelling all the energetics involved in the process of folding, and then in finding the structure with lowest free energy. This approach is based on the **thermodynamic hypothesis**, which states that the native structure of a protein is the one for which the free energy achieves the global minimum. While *ab initio* prediction is clearly the most difficult, it is arguably the most useful approach.

Semi-Empirical Methods

By contrast, experimental parameterization or neglect of the most time-consuming components of the Schrödinger equation gave rise to semi-empirical methods (partly from experimental data). Although not as reliable and versatile as *ab initio* techniques, such algorithms can be routinely applied to much larger systems.[7-8]

⬥ Semi-empirical methods use parameters that compensate for neglecting some of the time-consuming mathematical terms in Schrödinger's equation, whereas *ab initio* methods include all such terms.

⬦ The parameters used by semi-empirical methods can be derived from experimental measurements or by performing *ab initio* calculations on model systems, and are limited to hundreds of atoms.

⬦ They can be applied to organics, organo-metallics, and small oligomers (peptide, nucleotide, saccharide)

⬦ They can be used to study ground, transition, and excited states (certain methods).

⬦ Specific implementations include AMPAC, MOPAC, and ZINDO.

Quantum mechanics can satisfactorily predict the properties/reactivity connected with the electrons like bond formation and bond breaking (reactions) reactivity, based on HOMO-LUMO (Highest Occupied Molecular Orbital-Least Unoccupied Molecular Orbital), etc. The concept of HOMO and LUMO is useful in predicting the feasibility of a reaction and stereochemistry of the product formed.

But quantum mechanics works less efficiently for molecules like proteins which contain thousands of atoms, and their calculations are computationally demanding. The ability to characterize a molecule using quantum mechanics opens up a whole new range of descriptors [9,10] or molecular representations that can aid drug discovery. Many of these descriptors are beyond the reach of classical potentials and by their very nature can be used to gain a qualitative understanding of protein–ligand interactions [11] and then be used in the rational design of drug molecules. Linear-scaling QM methods have made therapeutically important protein targets routinely accessible to qualitative analysis. For example, new QM-derived descriptor classes, such as molecular electrostatic potential (ESP) maps (Figure 2.1), local hardness and softness, Fukui indices, [12] frontier orbital analysis, [13,14] density of states, etc. can be used to probe protein–ligand complexes. ESP maps have been widely used as tools for characterizing protein or DNA binding sites. [15]

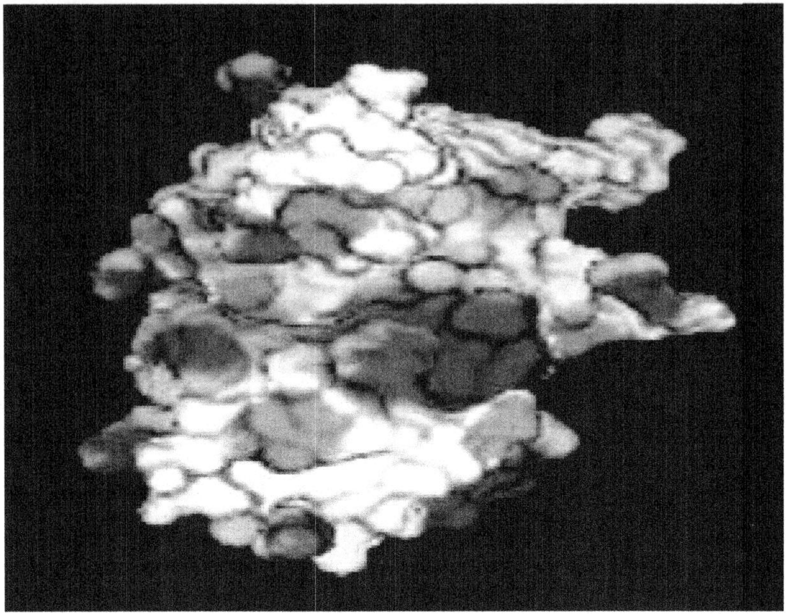

Figure 2.1 Electrostatic potential of an enzyme produced using DELPHI (*See* Plate 2a)

BORN-OPPENHEIMER APPROXIMATION

The Born-Oppenheimer approximation enables to solve the Schrödinger equation for more than one-electron systems. It separates electron and nuclear motion based on the idea that nuclear mass is so much larger than electron mass that the nuclei are basically "fixed" particles. Since the nuclei are much heavier than the electrons, any movement in the nuclei, correspondingly is quickly adjusted by the electrons. Thus the molecular wave function $\Psi(a,b)$ where a and b are the positions of the electrons and the nuclei, can be separated as $\Psi(a) * \Psi(b)$. The Schrödinger equation is solved only for the electrons in the nuclei field. For the nuclear potential (both rotational and vibrational) the electronic motion appears smeared.

Molecular Mechanics/Force-Field Approach

Molecular mechanics calculation is the fastest approach to obtaining data in computational modelling. The basis of molecular mechanics is the Born-Oppenheimer approximation. Molecular mechanics simulations use the laws of classical physics to predict the structures and properties of molecules. It is the protocol to relate the energy of a molecule with the geometric variables such as bond length, bond angle, torsional angle, etc. The molecule is treated at the atomic level, i.e., the electrons are not treated explicitly and this is the reason why unlike quantum mechanics, molecular mechanics cannot explain phenomena related to electrons. This includes reactivity of molecules, bond formation–breaking, etc.

Many different molecular mechanics methods exist, each characterized by a **force field**. Force field is a set of functions and parameterization used in molecular mechanics calculations. This also includes mapping of potential energy surfaces and simulation of protein–ligand docking. Its main application is, arriving at the energy minimum structure of a molecule using force-field and portrays, how the energy of a system varies with the conformation. Unlike quantum mechanical methods, it can be applied to a system containing thousands of atoms and can optimize the initial geometry constructed through molecular modelling (say homology modelling).

A general application of molecular mechanics (MM) depends on the validity of two underlying assumptions.

1. The total molecular energy of a system is separable into different energy contributions.

2. The different energy terms with their distinct parameters are transferable from the limited data they were parameterized for, to all molecular systems under consideration.

Energy Function

Molecular mechanics uses an energy function, which defines the different contributors to the energy of a molecule like bond stretching, bending, etc. (Figure 2.2) so that given a particular conformation (i.e., coordinates for all the atoms), the energy of the molecule can be calculated. The energy function is empirical and derived from either quantum mechanical calculations or experimental data. This energy function involves both bonded and non-bonded interactions. Bonding interactions include bond lengths, bond angles and torsional angles and are assumed to be harmonic (just like two balls attached to a spring).

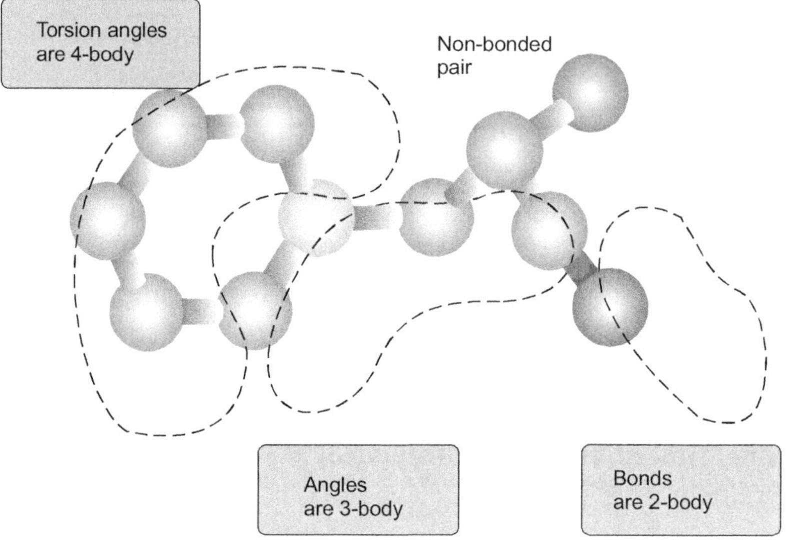

Figure 2.2 A schematic diagram portraying the different bonded interactions

Since spring action (Figure 2.3) obeys Hookes Law

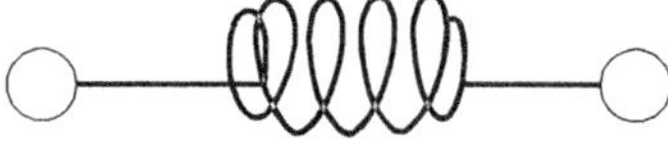

Energy contribution of the bond length (which can be equated to the stretching of the string), is given by

$$E_{str} = 1/2\,k_b(b - b_0)$$

where k_b is the force constant for stretching and b_0 is the bond length in the built structure and b is the original/experimental bond length.

Similarly, the energy analogs of the angle (bending) and torsion are given by

$$E_{\text{bend}} = 1/2\,k_\theta(\theta - \theta_0) \text{ and } E_{\text{tors}} = 1/2\,k_\varphi(1 + \cos(\varphi - \varphi_0))$$

Non-bonding interactions such as van der Waals and coulombic interaction are also defined.

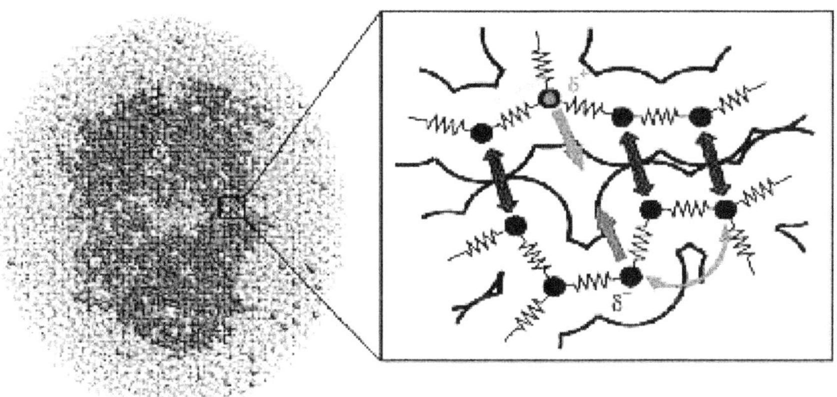

Figure 2.3   A depiction of spring like interactions in a macrocyclic system between the atoms including the dipoles (*See* Plate 3)

Force Field

Within the molecular mechanics approach, a set of potential functions, defining bond stretch, bond angle (valence and dihedral) and distortion energy of a molecule as compared with its non-strained conformation (standard values of bond lengths and angles) is termed as **force field**. Transferability is one of the key properties of a force field. A set of transferable empirical force constants is pre-assigned and the harmonic approximation is usually employed. These are normally derived from experimental techniques like X-ray crystallography.

Some force fields may contain terms for interactions between non-bonded atoms, electrostatic, hydrogen bond and other structural effects as well as account for observed vibrational frequencies of a given series of congeneric molecules. Force field also includes the types of atoms, which differentiates the hybridization state. An efficient force field would have many atom hybridization states and hence would include additional parameters corresponding to them. The force field used in molecular modelling is primarily designed to reproduce structural properties and also to predict other properties such as molecular spectra. Force fields are normally classified into different classes based on their efficiency. To describe a more complicated system which involves specialized interactions, higher class force fields are used.

Thus the contribution to the total energy of a molecule can be separated into

$$E = E_{\text{stretching}} + E_{\text{bending}} + E_{\text{van der Waals}} + E_{\text{electrostatic}}$$
$$+ E_{\text{hydrogen bond}} + \text{crossterms} \tag{1}$$

Equation 1 gives the general form of the expression used in force fields. The first four terms contribute to the deviation of the bond length, bond angle, torsional angle and non-bonded distances respectively from their ideal value (normally generalized from X-ray crystallographic data or quantum mechanical calculation). As mentioned earlier the direct contact of atoms are likened to springlike interactions. Most advanced force fields can include cross terms which are mixed movements such as stretch–bend, bend–bend, torsion–stretch, etc. The more applicable terms a force field includes, the wider becomes its application. The most commonly used protein force fields incorporate a relatively simple (sometimes called "Class I") potential energy function:

$$v(r) = \sum_{\text{bonds}} k_b (b - b_0)^2 + \sum_{\text{angles}} k_\theta (\theta - \theta_0)^2 + \sum_{\text{torsions}} k_\phi [\cos(n\phi + \delta) + 1]$$
$$+ \sum_{\substack{\text{nonbond} \\ \text{pairs}}} \left[\frac{q_i q_j}{r_{ij}} + \frac{A_{ij}}{r_{ij}^{12}} - \frac{C_{ij}}{r_{ij}^6} \right]$$

The first three summations are over bonds (1–2 interactions), angles (1–3 interactions), and torsions (1–4 interactions). Force constants and idealized bond lengths and angles are taken from crystal structures.

A Class II force field is designed to be a transferable force field for accurately treating conformational energetics and non-bonded interactions (e.g. OPLS force field). This would, ideally, produce a force field that was adequate for both gas phase and condensed phase calculations. Some force fields have additional terms in the above equation which adds to the superiority of the force field. The parameters are fit to reproduce experimental data.

Classical Force Fields

Force constants of the energy terms must be well characterized and be used appropriately for the molecule under investigation. To overcome the disadvantage, several different types of force fields have been developed. These include Assisted Model Building using Energy Refinement (AMBER) Chemistry at HARvard, Macromolecular mechanics (CHARMm), Constant Valence Forcefield (CVFF), Consistent Forcefield (CFF), Molecular Mechanics Allinger Force Field 2 (MM2) and Merck Molecular Force Field (MMFF). These force fields and others, streamline the computational process and enable modelling of larger structures such as enzymes and DNA which are used to help predict molecule interaction and structure.

Energy Minimization

Energy minimization techniques are used to iteratively refine the conformation of a molecule so as to minimize its energy. This is particularly useful for removing bad clashes between atoms that may have developed when building or modelling a molecule (e.g. a homology-modelled structure of a protein).

In the above structure, these are two flexible portions, namely, the side chains [dotted lines which can result in generating conformations] and the cyclohexane ring itself (which can adopt chair, boat or twist conformation).

But any conformation wherein the amine (–NH$_2$) and the hydroxyl (–OH) groups are in the same plane would be higher energy or unstable conformation. This is due to the repulsion between the lone pairs "Nitrogen" and "Oxygen".

The high energy would be due to the repulsion of the lone pair of electrons, which both the nitrogen and oxygen possess. Correspondingly, the least energy conformation would be the one in which both are *trans* to each other.

The steps in energy minimization include:

1. Building the molecule according to its bond connectivity

2. Applying appropriate force field

    e.g. Small molecule   cff91

           Proteins   cvff, AMBER

           Organometallics   ESFF

3. Fixing charges—if it is neutral molecule, the charge is zero. If there is a positively charged group like $\overset{+}{N}H_3$ or COO^- the overall charge is +1 and −1 respectively. The choice of minimization algorithm should be matched to the problem at hand. A good strategy is to switch minimizers as the calculations progress, so that each algorithm operates in the regime for which it was designed.

As Figure 2.4 depicts, when a particular model/chemical structure is built, it would be in a high energy conformation (starting energy), since no energy criteria is used while building a model. Just like a ball rolling down the hill, the lesser energy (local minimum) state is attained by using minimization algorithm. But this is only a local minima, but efficient algorithms will locate the global minima which is the one with the least energy conformation. The plot given below is a graphical depiction of how energy varies with change in the dihedral angle. Both the x- and y-axes would have values

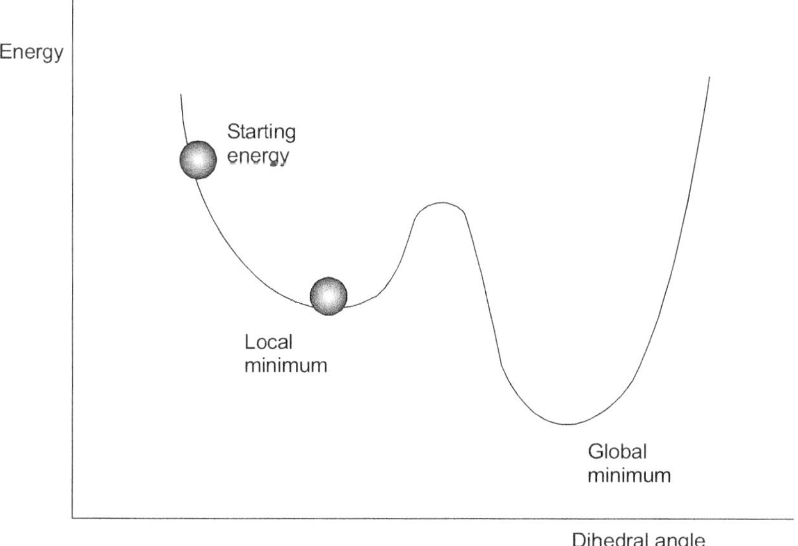

Figure 2.4 The process of energy minimization

A molecule might have more than one least energy conformation and one of them would correspond to the crystallographically obtained one. That is the reason why crystal structures are starting point in drug design, needing no energy minimization.

Algorithms for Energy Minimization

Simplex

This is not a gradient minimization method. Used mainly for very crude, high energy starting structures.

Steepest Descent

This works well (convergence is faster) when the built structure is far away from the minima, i.e., if the molecule has been built from the scratch. This can lead to backtracking. Backtracking is a phenomena in which the identified least energy conformation is re-identified due to the reversal of path.

Conjugate gradients or Powell

Remembers the gradients calculated from previous steps to help reduce backtracking. Generally finds a minimum in fewer steps than Steepest Descent. May encounter problems when the initial conformation is far from a minimum.

Newton–Raphson or BFGS

Predicts the location of a minimum, and heads in that direction. May find a minimum in fewer steps than the gradient-only methods. May encounter serious problems when the initial conformation is far from a minimum.

To sum up, molecular mechanics methods are based on the following principles:

- ◈ Nuclei and electrons are lumped into atom-like particles.

- ◈ Atom-like particles are spherical (radii obtained from measurements or theory) and have a net charge (obtained from theory).

- ◈ Interactions are based on springs and classical potentials.

◈ Interactions must be pre-assigned to specific sets of atoms.

◈ Interactions determine the spatial distribution of atom-like particles and their energies.

Molecular mechanics force field can be used to calculate

◈ The conformation with the least energy

◈ The relative energy of different conformations of a molecule

◈ Vibrational frequencies

◈ Dipole moments

◈ Intermolecular interactions

This also includes areas where phenomena involving electrons are involved like electronic transitions (e.g. fluorescence), electron transport phenomena (e.g. Krebs cycle, redox reactions), Proton transfer (e.g. acid base transfer), etc.

Advantages and Limitations of the Force-Field Approach

Advantages

◈ Speed of calculations

◈ Large systems can be treated (several thousands atoms with a PC)

◈ Easy to include solvent effects and crystal packing

Limitations

◈ Lack of good parameters (for molecules which are out of the ordinary)

◈ The predicting power is very limited

◈ Transferability is limited

◈ Cannot simulate breaking and formation of bond

Implication of Molecular Mechanics in Drug Design

The aim of creating molecular models towards drug design is to mimic the natural processes as close as possible. And molecules interact in their least energy conformation. Hence these models would be the ideal point for the different protocols in drug design. But the least energy conformation need not be the most active conformation.

CASE STUDY

Energy Minimization of 2-Acetoxy Cyclohexanol using ArgusLab and VegaZZ

ArgusLab

ArgusLab is a tool which includes, spectroscopy, graphics and visualization, drug-docking, and high-level *ab initio* calculations. It is available at http://www.arguslab.com

VegaZZ

For this case study VegaZZ a free program for molecular mechanics would be used. This can be downloaded from www.ddl.unimi.it after registration.

The chemical structure of the molecule whose structure is to be minimized is given below

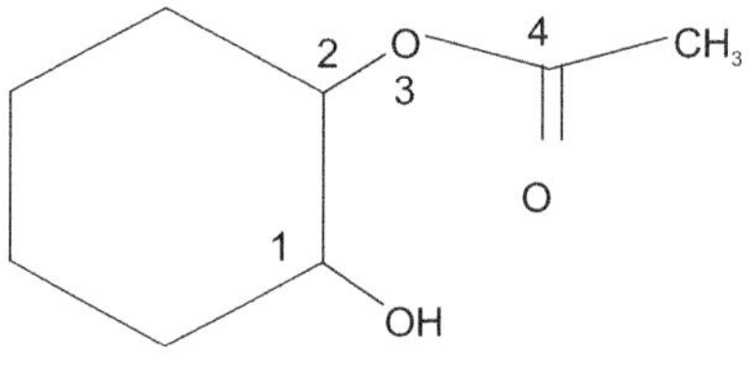

2-Acetoxy cyclohexanol

How to build the molecule?

Given below is the snapshot of the VegaZZ screen

On the left side, clicking the yellow box (last but one from down), would result in the appearance of **Add fragment/s** box

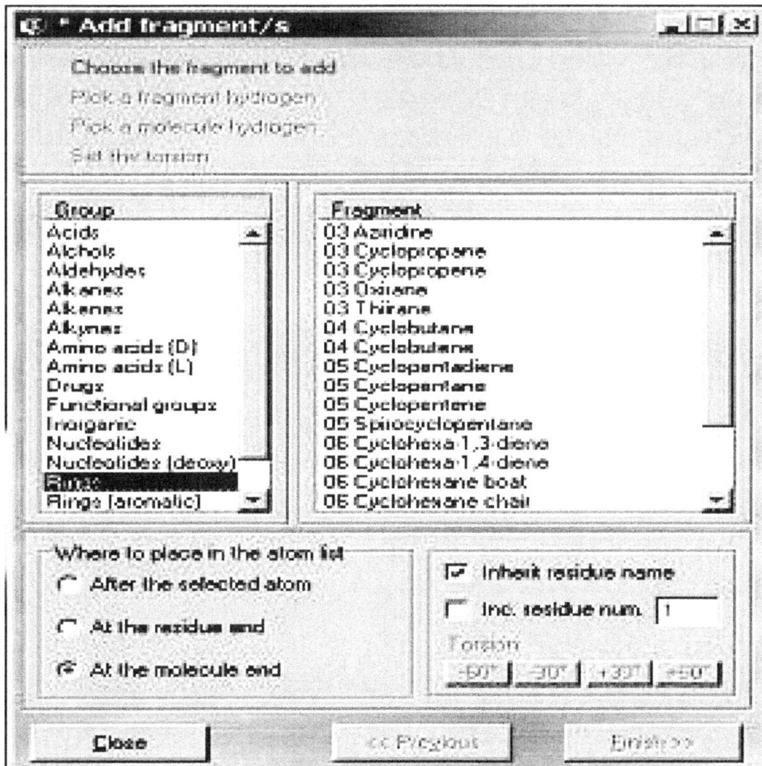

from which **Rings→Cyclohexane chair**, can be selected, which would appear on the screen.

Using the **Edit** menu, select the **Add** submenu and select the **Atoms**, which would result in the following pop-up. Using this, atoms can be added/removed, the bond order/ hybridization can be altered, etc.

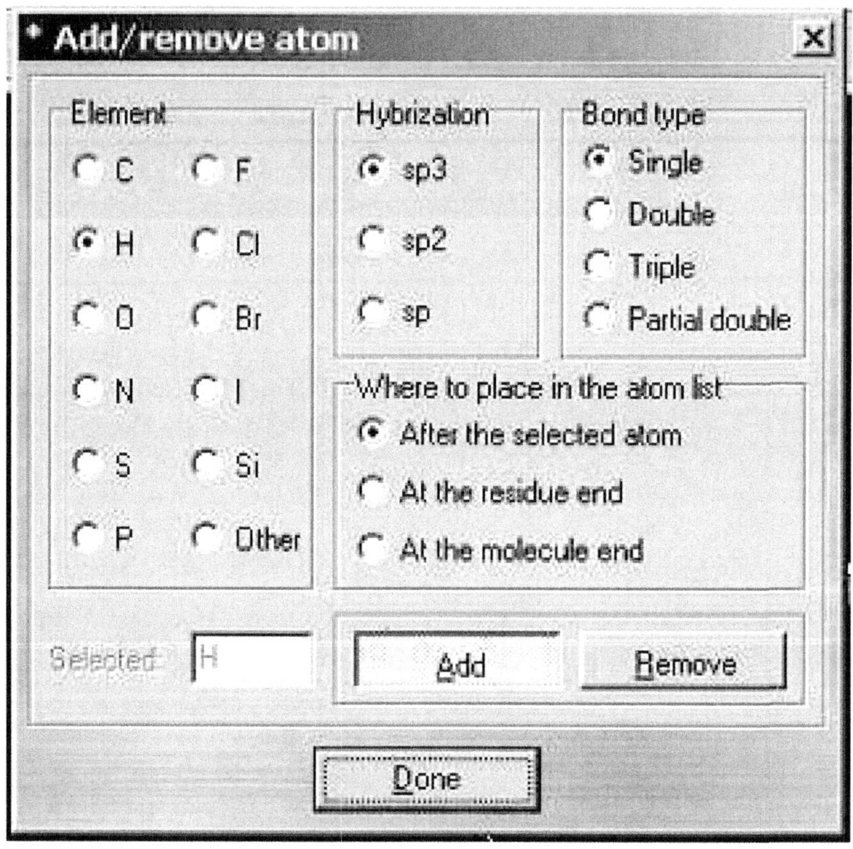

After building the complete molecule (with the side-chain substituents), right-click the mouse, select **Measure→Torsion** and choose the torsion $C_1-C_2-O_1-C_7$.

The minimization would be done, based on the torsion angle (i.e., how the energy varies with torsion angle). Since towards the end of the side chain the rotational degree of freedom increases, the torsion near the ring C_2-O_1 has been chosen

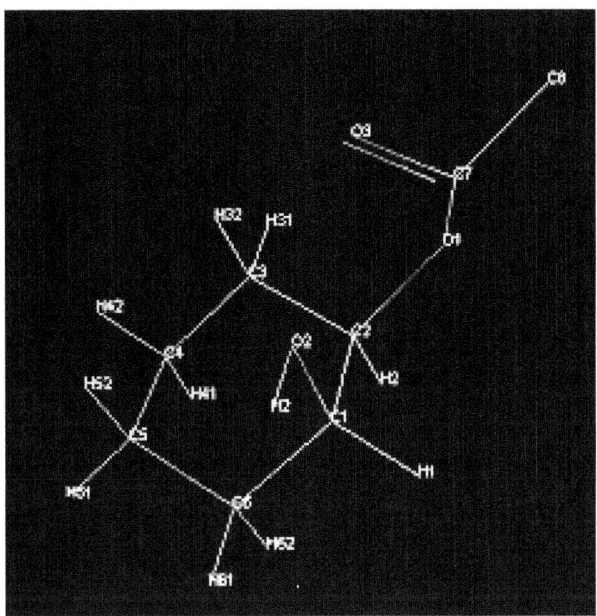

(selecting the end torsions like O_4-C_7 would result in multiple minima). The force field used is the TRIPO S and the Gasteiger charges are fixed.

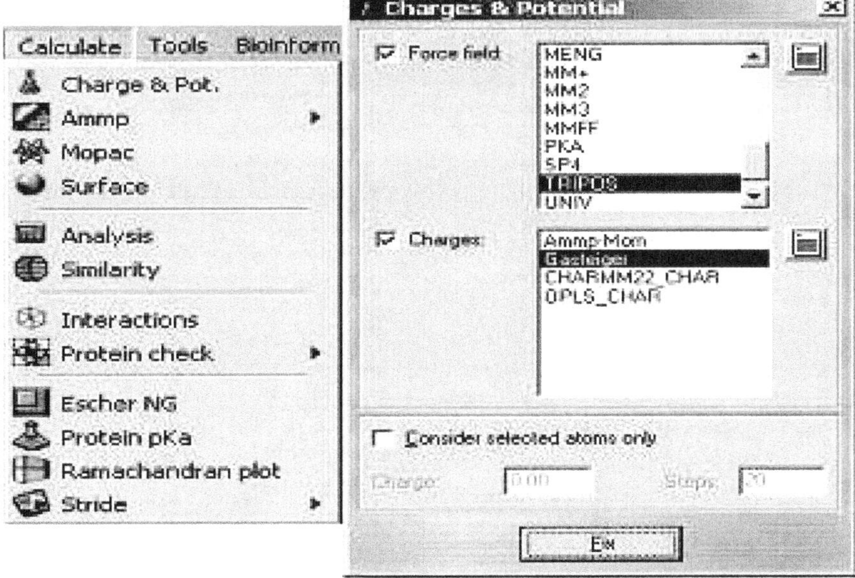

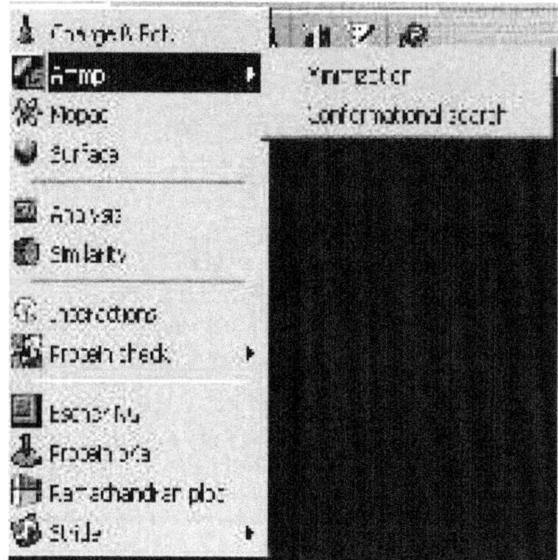

Steepest descent, Conjugate gradient, Quasi-Newton and others are all algorithms for minimization. We shall choose conjugate gradient.

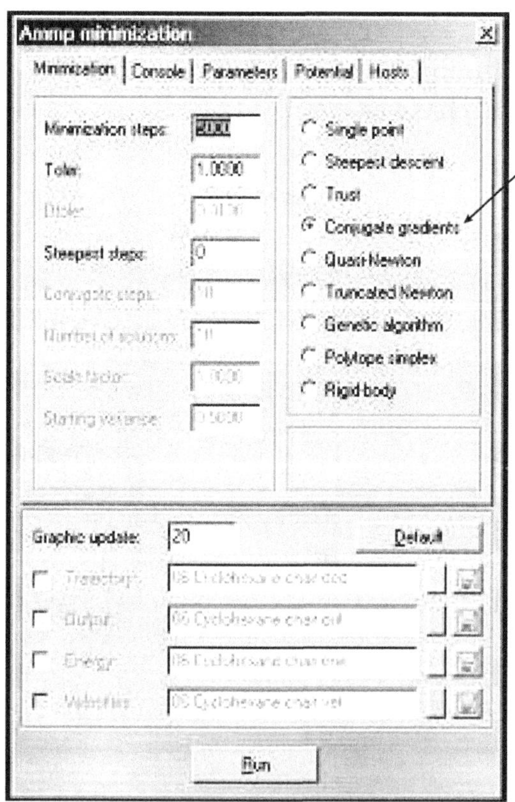

Set the parameters as given below

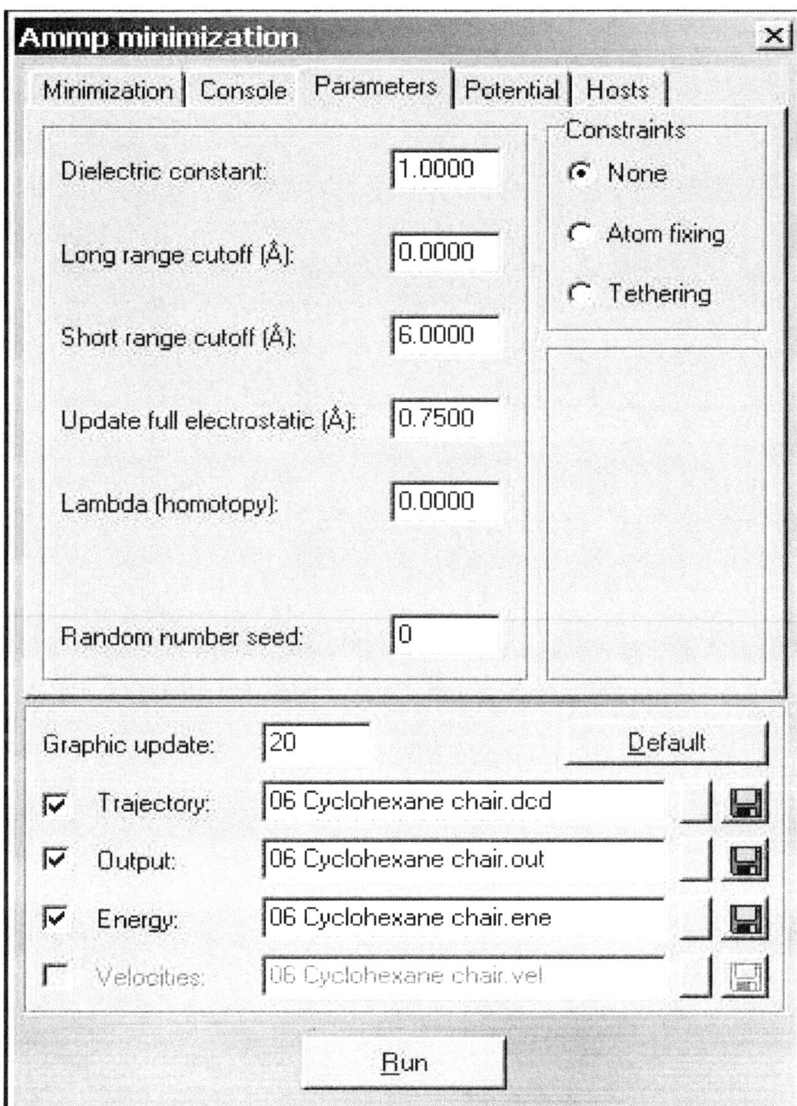

Executing the **Run** command will result in minimization and there would be a change in selected torsion angle. The resulting conformation would correspond to one of the minimum energy one. Given below are the diagrams of the structure before and after minimization.

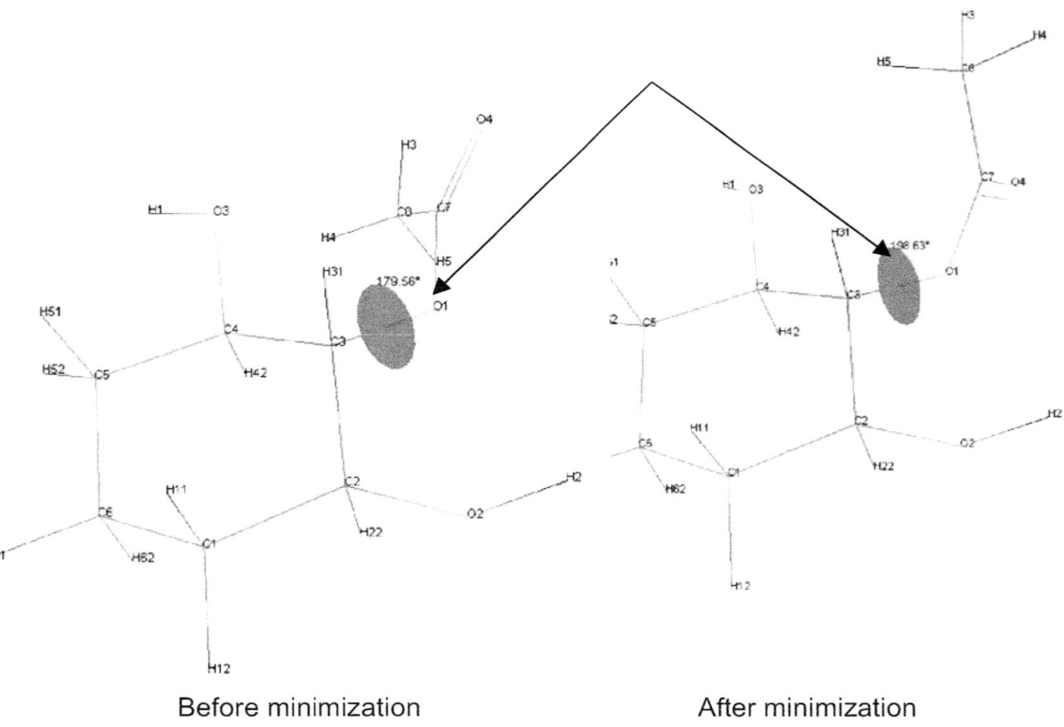

Figure 2.5   The change in torsional angle before and after minimization

As we can clearly see, the torsion angle in the un-minimized/built structure is 179.56° and after minimization it is 198.63°, with the intra-molecular high-energy interactions relieved in the process. The output files *.dcd and *.out are the trajectory and energy profile files respectively.

MOLECULAR DYNAMICS

The motion of molecules is important in essentially all chemical interactions and is of particular interest in biochemistry. For example, the binding of substrates to enzymes, the binding of antigens to antibodies, the binding of regulatory proteins to DNA, and the mechanisms of enzyme catalysis are enhanced and sometimes completely determined by the conformational flexibility of the molecules. Different domains of an enzyme can have very different motional freedom. The problem of protein folding is the determination of the trajectory of the macromolecule as it assumes its active conformation after or during protein synthesis.

Molecular dynamic simulations are computational tools used to estimate equilibrium and dynamic properties of complex systems that cannot be calculated mathematically by hand. Molecular dynamic simulations investigate the time development of a many-particle system and are evaluated by numerically integrating Newton's equations of motion. The trajectory is determined by integrating Newton's equations of motion for the bond stretching, angle bending, and dihedral torsions of the molecule. Since molecules are always in motion, the minimization process computes the forces on the atoms and adjusts their positions to minimize repulsive interactions. To be concise, *molecular dynamics calculates the forces and moves atoms in response to the forces.*[16] Forces are generated by atom–atom interactions, which are usually given in terms of an empirical potential. Parameters for the potential are usually obtained by fitting to either *ab initio* calculations or experimental data. A typical empirical force field formulation of the potential energy for N atoms at positions $r_1 \dots r_N$, can be written as

$$E = \sum_{bonds} \frac{a_i}{2}(l_i - l_{i0})^2 + \sum_{angles} \frac{b_i}{2}(\theta_i - \theta_{i0})^2 + \sum_{torsions} \frac{V_n}{2}(1 + \cos(n\omega - \gamma))$$

$$+ \frac{1}{2}\sum_{i=1}^{N}\sum_{j \neq i}^{N} 4\epsilon_{ij}\left[\left(\frac{\sigma_{ij}}{r_{ij}}\right)^{12} - \left(\frac{\sigma_{ij}}{r_{ij}}\right)^{6}\right] + \frac{1}{2}\sum_{i=1}^{N}\sum_{j \neq i}^{N} \frac{q_i q_j}{r_{ij}}$$

By following the motions of a molecular system in space and time, a rich amount of information about the structural and dynamic properties of a system can be obtained.

Molecular dynamics simulations, of both classical and quantum nature, have proven to be invaluable in elucidating the structural, mechanical and chemical properties of diverse sets of materials. For example, MD simulations have been used to study the liquid state/the bulk solid,[17] diffusion,[18] wetting phenomena, phase transformations,[19] and protein dynamics,[20] to name several examples.

The application of bio-molecular simulation can be categorized into three types:

1. Using simulation to bring bio-molecular structures alive, giving insights into the natural dynamics on different timescales of bio-molecules in solution.

2. To obtain thermal averages of molecular properties. This is used to calculate, for example, the bulk properties of fluids and the free energy differences for chemical processes such as ligand binding.

3. Thirdly, molecular dynamics can explore which conformations of a molecule or a complex are thermally accessible. This technique is used for exploring conformational space, for instance, in ligand-docking applications.[21]

Comparison of simulation and experimental data serves in assessing the computational methods and helps in further improving its accuracy.

Molecular dynamics simulations explore,

◈ Protein stability

◈ Conformational changes

◈ Protein folding

◈ Molecular recognition: proteins, DNA, membranes, complexes

◈ Ion transport in biological systems

The input for a MD simulation is an initial set of coordinates, obtained from NMR studies or X-ray diffraction analysis. The computational time required for a MD simulation grows with the square of the number of atoms in the system, because of the non-bonded interactions defined in the potential energy function, for example, Lennard–Jones potential

$$V(r) = 4\varepsilon \left[\left(\frac{\sigma}{r} \right)^{12} - \left(\frac{\sigma}{r} \right)^{6} \right]$$

(where ε is the depth of the potential well and r is the (finite) distance at which the interparticle potential is zero) or the Coulomb potential

$$E = \frac{q_i q_j}{4\pi\varepsilon_0 \varepsilon_r r_{ij}}.$$

The main advantage of molecular dynamics is the incorporation of the solvent environment (e.g. water). This is of main consequence because the proteins in the human system do exist in aqueous-based system along with other metabolites.

Performing longer simulations for larger systems does not guarantee the production of useful results. Key underlying issues for accuracy are adequate configurational sampling and the quality of the description of the intramolecular and intermolecular energetics.

Majority of the simulations involve solvent (aqueous) environment since most of the proteins exist in such systems.

Algorithms to Integrate the Equations of Motion

There are different algorithms used in integrating the equations of motion.

Verlet Leap Frog algorithm

The Verlet Leap Frog algorithm uses positions(r) and accelerations (a) at time t and the positions from time $t-\Delta t$ to calculate new positions at time $t+\Delta t$.

$$v(t+\Delta t/2) = v(t-1/2\,\Delta t) + \Delta ta(t), r(t+\Delta t) = r(t) + \Delta tv(t+1/2\,\Delta t), a(t+\Delta t)$$
$$= f\left(t+\Delta t\right)/m$$

where $f(t+\Delta t)$ (force) is evaluated from $-dv/dr$ at $r(t+\Delta t)\Delta t$ and v is the velocity.

The important criteria in the algorithms are the time space Δt, which depends on the problem that is being worked on as well as the integrators. It should be chosen such that the velocities and accelerations are constant over the time step used.

Beeman algorithm

$$r(t+\delta t) = r(t) + v(t)\delta t + \frac{2}{3}a(t)\delta t^2 - \frac{1}{6}a(t-\delta t)\delta t^2$$

$$v(t+\delta t) = v(t) + v(t)\delta t + \frac{1}{3}a(t)\delta t + \frac{5}{6}a(t)\delta t - \frac{1}{6}a(t-\delta t)\delta t$$

Approximations in Molecular Dynamics

Periodic boundary conditions (PBC)

Periodic boundary conditions enable a simulation to be performed using a relatively small number of particles in such a way that the particles experience forces as though they were in a bulk solution.

In this approach, the original box containing a solute and solvent molecules is surrounded with identical images of itself, i.e., the positions and velocities of corresponding particles in all of the boxes are identical. The common approach is to use a cubic or rectangular parallelepiped box, but other shapes are also possible (e.g. truncated octahedron). By using this approach, we obtain what is in effect an infinite-sized system.

Stochastic boundary conditions (SBC)

Stochastic boundary conditions allow reducing the size of the system by partitioning the system into essentially two portions: a **reaction zone** and a **reservoir region**. The reaction zone is that portion of the system which needs to be studied and the reservoir region contains the portion which is inert and uninteresting. For example, in the case of an enzyme, the reaction zone should include the proximity of the active centre, i.e., portions of protein, substrate, and molecules of solvent adjacent to the active centre. The reservoir region is excluded from molecular dynamics calculations and is replaced by random forces whose mean corresponds to the temperature and pressure in the system. The reaction zone is then subdivided into a **reaction region** and a **buffer region**. The stochastic forces are only applied to atoms of the buffer region, in other words, the buffer region acts as a "heat buffer".

RUNNING MOLECULAR DYNAMICS

1. The initial coordinates are taken from crystallographic structure or the modelled structure or even a protein–protein complex.

2. The MD simulation is run using a package like GROMACS (www.gromacs.org/) which is a free tool.

3. The pH has to be specified so that appropriate charges can be fixed (for example using GROMOS96).

4. The object of study is placed in an environment like water in a defined grid box. Many water models such as simple extended point charge (SPC/E), TIP4P , TIP5P, etc. are used.

5. Thermodynamic properties such as temperature, volume, and pressure are used to describe a large system of particles. Such a system and all its thermodynamic properties form an ensemble– NVTP (number of particles N, volume V, temperature T, pressure P).

6. Short-range electrostatic van der Waals forces and electrostatic interactions are computed.

7. The energy of the macromolecule/object of study is minimized in the environment the temperature variation (Kelvin) is set up along with the increment steps.

8. After the completion of the simulations, the results are analysed.

LIMITATIONS OF MOLECULAR DYNAMICS

The limitations to molecular dynamics are similar to molecular mechanics. Unfortunately, the simulations ignore the electronic motion and quantum effects thus taking into account only the movement of an atom's nuclei. Although this type of simulation is useful for a wide range of simulations, it is unsuitable for reactions involving bond formation and cleavage, polarization, and chemical bonding of metal ions. Moreover, it is also unsuitable for low temperatures due to the formation of large energy gaps as dictated by quantum physics, which is much larger than the thermal energy available to the system. The system is confined to one or a few discrete low-energy states under the low-temperature conditions.

SIMULATED ANNEALING

Simulated annealing[22] is a special case of either Molecular Dynamics ('quenched' MD) or Monte Carlo simulations, in which the temperature is gradually reduced during the simulation. Often, the system is first heated and then cooled. Thus, the system is given the opportunity to surmount energetic barriers in a search for conformations with energies lower than the local-minimum energy found by energy minimization. This improved equilibration can lead to more realistic simulations of dynamics at low temperature. The simulated annealing method is based on the similarity that exists between locating the global minimum of the potential energy function of a molecule and the slow cooling required for obtaining a perfect crystal. This is true when crystallization compounds where slow crystallization (in terms of time) yield better crystals with less mosaicity.

Application of this concept to the exploration of the conformational space can be translated in terms of starting the simulation at sufficiently high temperature and subsequently decreasing it gradually until the system is frozen in the global minimum.

MOLECULAR DYNAMICS IN DRUG DESIGN

The need to account for the dynamic behaviour of a receptor has long been recognized as a complicating factor in computational drug design. The use of a single, rigid protein structure—usually from a high-quality X-ray crystal structure—still is the standard in most applications. The choice to use a single protein structure is usually based on speed. For example, if a large database of compounds is to be screened for binding affinity, several conformers of each compound will be compared with each protein configuration. Although it is more accurate to use many representative protein configurations, it quickly becomes a very slow process to screen each conformer of each ligand against each protein configuration. It is usually impractical to attempt such prohibitively slow calculations because there is an unfortunate but necessary trade-off between speed and accuracy in computer modelling.

CASE STUDY

The following case study[23] illustrates the utility and application of molecular dynamics.

The complex between Azurin (AZ), an electron carrier (Azurin is one of a class of copper-containing proteins called blue-copper proteins, so called because of their striking blue colour) in the respiratory chain of *Pseudomonas aeruginosa* and the haeme protein cytochrome (C551), involved in the de-nitrification process.

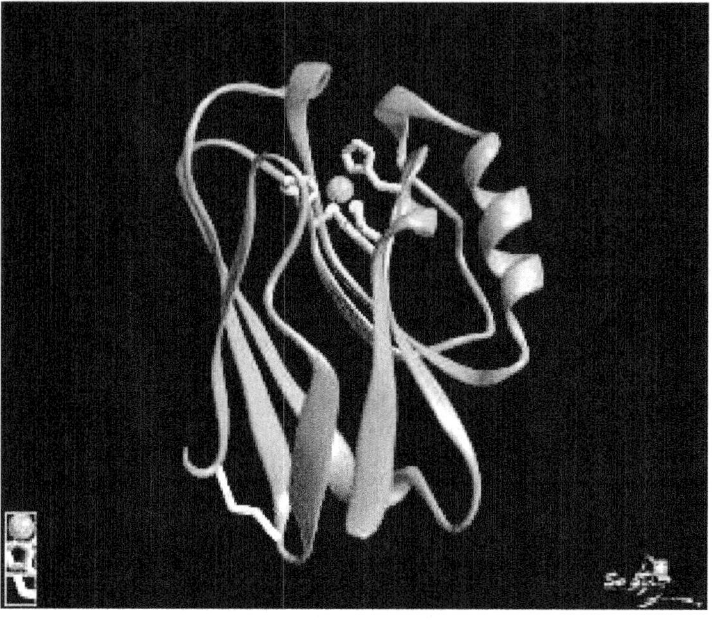

(*See* Plate 4)

In vitro studies have shown that AZ exchanges electrons with C551. Furthermore, atomic force spectroscopy experiments have probed the specificity and the bio-recognition between these two protein partners. However, since this complex is not accessible to crystallographic investigation, most likely due to its limited lifetime, its real spatial arrangement is still missing.

A molecular dynamics simulation, carried out for 60 ns at full hydration, showed that the two proteins substantially maintain their initial relative arrangement upon forming the complex. An analysis of the complex structures, sampled during the dynamics, has revealed that the electron-transfer (ET) properties change in time with respect to involvement of different residues and the occasional participation of a few water molecules, suggesting a dynamically assisted ET process. Furthermore, the metal-to-metal distance monitored as a function of time is characterized by fast fluctuations around a value of 1.6 nm, with periodic jumps to a value of 2 nm about every 15 ns; such a behaviour, indicative of nonlinear dynamics, has suggested some implications to the ET properties of the complex.

REFERENCES

1. Neil, R. Taylor and Mark von Itzstein. (1994). "Molecular modeling studies on ligand binding to sialidase from influenza virus and the mechanism of catalysis." *J. Med. Chem.* 37:616–624.

2. Hu, R., Chung, T.C., Chadwick, P.C., Bridgwater, J., Neria, E. and Karplus, M. (1997). "Molecular dynamics of an enzyme reaction: proton transfer in TIM." *Chemical Physics Letters.* Vol. 267, Number 1, 14 March. pp. 23–30(8).

3. Sikorski, A. and Skolnick Monte Carlo, J. "Simulation of equilibrium globular protein. folding: alpha-helical bundles with long loops." *Proc. Natl. Acad. Sci. USA.* 86(8): 2668–2672.

4. Hehre, W.J. *et al.* (1986). *Ab initio Molecular Orbital Theory.* John Wiley & Sons.

5. Ragavachari, K. and Curtiss, L.A. (1995). "Evaluation of bond energies to chemical accuracy by quantum chemical techniques." In: *Modern Electronic Structure Theory.* (Yarkony, D.R. ed.). World Scientific Press. pp. 991–1012.

6. Curtiss, L.A. and Ragavachari, K. (1995). "Calculation of accurate bond energies, electron affinities, and ionization energies." In: *Quantum Mechanical Electronic Structure Calculations with Chemical Accuracy: Understanding Chemical Reactivity.* (Langhoff, S.R. ed.). Kluwer Academic Press. pp. 139–171.

7. Dewar, M.J. and Thiel, W. (1977). "Ground states of molecules. 38. The MNDO method-Approximations and parameters." *J. Am. Chem. Soc.* 99: 4899–4907.

8. Stewart, J.J.P. (1989). "Optimization of parameters for semi empirical methods. 2. Applications." *J. Comput. Chem.* 10: 221–264.

9. Khandogin, J. and York, D.M. (2004). "Quantum descriptors for biological macromolecules from linear-scaling electronic structure methods." *Proteins Struct. Funct. Bioinform.* 56: 724–737.

10. Kenny, B., Lipkowitz and Donald, B. Boyd. (eds.). (2003). *Reviews in Computational Chemistry.* John Wiley and Sons, Inc. Vol. 18.

11. Ulrich Nienhaus, G. (2005). *Protein-Ligand Interactions: Methods and Applications.* Humana Press.

12. Mineva, T., Parvanov, I., Petrov, Neshev, N. and Russo, N. Fukii. (2001). "Indices from perturbed kohn-sham orbitals and regional softness from mayer atomic valences." *J. Phys. Chem. A.* 105 (10): 1959–1967.

13. Ian Fleming. (1976). *Frontier Orbitals and Organic Chemical Reactions.* Wiley.

14. Sean, E. O'Brien, Helen, L. Browne, Tracey, D. Bradshaw, Andrew, D. Westwell, Malcolm, F.G. Stevens and Charles, A. Laughton. (2003). "Anti tumor benzothiazoles. Frontier molecular orbital analysis predicts bioactivation of 2-(4-aminophenyl) benzothiazoles to reactive intermediates by cytochrome P4501A1." *Org. Biomol. Chem.* 1: 493–497.

15. Celeste Sagui and Thomas, A. Darden (1999). "Molecular dynamics simulations of biomolecules: long-range electrostatic effects." *Annu. Rev. Biophys. Biomol. Struct.* 28: 155–79.

16. Tomas Hansson, Chris Oostenbrink and Wilfred, F. van Gunsteren. (2002). "Molecular dynamics simulations." *Current Opinion in Structural Biology.* 12:190–196.

17. Noël Jakse and Alain Pasturel. (2005). "Molecular-dynamics study of liquid nickel above and below the melting point." *J. Chem. Phys.* 123: 244512.

18. Shimizu, A. and Tachikawa, H. (2003). "Molecular dynamics simulation on diffusion of lithium atom pair in $C_{150}H_{30}$ cluster model for glassy carbon at very low temperatures." *Electrochimica Acta.* Vol. 48, Number 12, pp. 1727–1733.

19. Mar, W., Hautman, J., Klein. (1995). "Molecular dynamics studies of microscopic wetting phenomena on self-assembled monolayers, M.L." *Computational Materials Science.* Vol. 3, Number 4, pp. 481–497(17).

20. Cheong, W.C.D. and Zhang L.C. (2000). "Molecular dynamics simulation of phase transformations in silicon monocrystals due to nano-indentation." *Nanotechnology.* Vol.11, No. 3. pp. 173–180.

21. Anwar Rayan, Hanoch Senderowitz, and Amiram Goldblum. (2004). "Exploring the conformational space of cyclic peptides by a stochastic search method." *Journal of Molecular Graphics and Modelling.* Vol. 22, Issue 5, pp. 319–333.

22. Kirkpatrick, S. and Gelatt, C.D. Jr. (1983). "Optimization by simulated annealing." *Science.* 220: 671– 680.

23. Bizzari, A.R., Brunori, E., Bonanni, B. and Cannistaro, S. (2007). "Docking and molecular dynamics simulation of the azurin–cytochrome c551 electron transfer complex." *J. Mol. Recognit.* 20: 122–131.

Freeware for Molecular Mechanics and Dynamics

TINKER-http://dasher.wustl.edu/tinker/

GChemical-http://www.bioinformatics.org/ghemical/ghemical/index.html

GROMACS-http://www.gromacs.org

Argus Lab-http://www.planaria-software.com/

FungiMol—http://www.infoscreen.com/fungimol

VMD-http://www.ks.uiuc.edu/Research/vmd/

Recommended Books for Reading

1. *Principles of Molecular Mechanics,* by Katsunosuke Machida, John Wiley & Sons, Inc.

2. *Molecular Mechanics Across Chemistry,* by A.K. Rappé, C.J. Casewit -1997 – University Science Books Sausalito.

3. *Molecular Dynamics Simulation: Elementary Methods,* by J.M. Haile Cambridge University Press.

Journals

1. *Journal of Molecular Structure: THEOCHEM,* Elsevier Science.

2. *Journal of Molecular Graphics and Modelling,* Elsevier Science.

3. *Journal of Chemical information and Modeling,* ACS.

4. *Journal of Medicinal Chemistry.*

REVIEW QUESTIONS

1. Which of the methods—nuclear magnetic resonance spectroscopy or X-ray diffraction studies—would you choose for elucidating the structure of the following compound. Justify your answer.

2. Explain the ADMET properties that are expected out of a chemical that is expected to exhibit drug-like properties.

3. What would be the choice of experiment in identifying the binding site in a protein? Briefly explain the protocol.

4. Explain the terms conformation and configuration and illustrate the role of configuration in the activity of a drug.

Do the compounds given below, differ in conformation or configuration?

5. Explain the classification of structural databases.

3

DRUG DESIGN: THE PROTOCOL AND METHODOLOGY

INTRODUCTION

Computer-aided techniques for the efficient identification and optimization of novel molecules with a desired biological activity have become a part of the drug discovery process. Such an approach could lead to a reduction in the cost of drug design and development by up to 50%.[1,2]

The traditional or the experimental process in drug discovery involves the following.

1. Target identification: This is the basic and important step in the birth of a drug. Target identification is to confirm that a particular enzyme/protein/DNA is the entity to be targeted in order to control the specified disease. This includes, microbe-based screening systems, functional assay screens, etc.

2. Target validation is carried out to confirm that the proposed macromolecule is involved and controls the process concerned.

3. Lead validation—involves screening of natural products or selection of leads through chemist's intuition.

4. Lead optimization generates hits, by chemical modification of the lead, and generate their bioactivity profile (by assaying it against a particular disease/ organism).

5. Drug developement: The lead compound is tested in the lab (*in vitro*) and on animals (*in vivo*) and their effect on biological systems is evaluated. Assessment of toxicology, pharmacodynamics and pharmacokinetics from the preclinical research activities. Clinicals trials have to follow regulations and guidelines from Food and Drug Adminstration (FDA) or other such agencies. The most active one is then identified and processed for further trials including three-phased clinical trial and approval by FDA. The Food and Drug Administration (FDA) is an agency of the United States Department of Health and Human Services and is responsible for the safety regulation of most types of

foods, dietary supplements, drugs, vaccines, biological medical products, blood products, medical devices, radiation-emitting devices, veterinary products, and cosmetics. This takes an appreciable amount of time (~12 years) with a price tag of 800 million US dollars.[3]

Drug discovery is a complex, multistage process that spans many scientific disciplines, including biology, chemistry, drug metabolism and clinical observation. There are two principal components in drug design, **the target** and the **drug**. A target may be considered as a molecular structure (chemically definable at least by a molecular mass) that will undergo a specific interaction with chemicals (drugs) because they are administered to treat or diagnose a disease. Structural knowledge of proteins and their ligands has aided in improving drug potency and selectivity. The interaction has a connection with the clinical effect(s).[4] Comparative analysis of biological sequence data, also referred to as genomic filtering, is a common and very effective method for identifying novel potential drug target proteins.[5,6]

In drug design, the main problem is in predicting reasonably which molecules from a pool of millions of possible compounds (compound libraries) will interact with a target of medical or biological interest. One approach is to utilize 3D models of target proteins and ligands. In order to build reasonable and useful models, as much information as possible has to be incorporated into the modelling process.

G-PROTEIN-COUPLED RECEPTORS—A COMPETITIVE TARGET TO PROTEINS?

The super family of G-protein-coupled receptors (GPCRs) forms the largest class of cell surface receptors. They are activated by a wide range of extracellular signals (such as small bioorganic amines, large protein hormones, neuropeptides, light, etc.) and transduce their signals across the plasma membrane, triggering a cascade of intracellular events leading eventually to the physiological response of the cell to the stimulus. The receptors have been linked to regulation of behaviour and mood through GPCRs found in the brain, to inflammation through receptors in the immune system, and to metabolic processes, among others. GPCRs represent one of the most important families of drug targets in pharmaceutical development and more than 20% of the top 200 best-selling prescription drugs interact with GPCRs. It is estimated that more than 40% of existing drugs work by targeting GPCRs, including compounds that target histamine receptors, such as Claritin for allergies or Pepcid for ulcers; angiotension receptors, such as Cozaar for blood pressure; or serotonin receptors such as the migraine drug Imitrex. While there are over 170 distinct subtypes of GPCRs with known ligands, over 200 human-derived GPCRs still remain "orphans" with no identified natural ligands and functions. Various pharmaceutical drug researches are being carried out to find out novel ligands, functions and patho-physiological mechanisms of such orphan GPCRs.

GLIDA is a GPCR-Ligand Database. GLIDA has advantages in that users can easily retrieve combined information about GPCRs, ligands and their relationship by either GPCR search or ligand search, providing more convenience in GPCR-related drug discovery. GLIDA can be accessed through http://pharminfo.pharm.kyoto-u.ac.jp/services/glida/

To apply structure-based drug design, it is necessary to have access to structural information of the drug target. Around 50,000 high-resolution structures have been deposited in the protein data bank; the majority of these however are on soluble proteins. In contrast, some 70% of current drugs are targeted against membrane proteins, for which only more than 100 structures are available.[7] This discrepancy relates mainly to the topological composition of membrane proteins. A large number of them such as G-protein-coupled receptors (GPCRs) and ion channels have a topology, including multi-spanning trans-membrane domains. There are thought to be more than 500 therapeutically relevant GPCRs out of a total of over 700 identified to date, although only one, rhodopsin, has been the subject of a full 3D X-ray crystallography study.

Two structurally related proteins, bacteriorhodopsin and sensory rhodopsin, which are not GPCRs but are part of the seven-helix membrane receptor family, have also been the subject of X-ray crystallographic studies and have been used in GPCR modelling studies.

Drugs/Ligands

An effective drug should

1. Be facile and economical to produce and deliver

2. Display favourable absorption, distribution, metabolism, excretion, and toxicity (ADMET) characteristics

3. Treat the targeted disease with specificity and efficacy.

Traditional medicines, as with other natural products, can offer powerful leads for therapeutic development because (unlike synthetic libraries) they already have documented effects on the organism. Compounds derived from medicinal extracts are appealing for several reasons.[8] They are often stereochemically complex, multi- or macrocyclic molecules with limited likelihood of prior chemical synthesis, and they tend to have interesting biological properties.

To quote a few compounds from plant source and possessing medicinal properties:

⬥ Artemisinin is a drug isolated from the Chinese herb *Artemisia annua* and used for treating malaria.

⬥ Triptolide is a potent immunosuppressive compound isolated from *Tripterygium wilfordii* and has been reported to inhibit autoimmunity, allograft rejection, and graft-versus-host diseases.

⬥ Celastrol, also an active compound of the root bark of *Tripterygium wilfordii*, has been used for years as a natural remedy for inflammatory conditions and also reported for its anti-tumour activity.

◈ Capsaicin is the active component of chilli peppers, which are plants belonging to the genus *Capsicum* and used as numbing or pain-reducing agent.

◈ Curcumin—isolated from the Indian curry spice turmeric possesses anti-tumour, antioxidant, anti-arthritic, anti-amyloid and anti-inflammatory properties.

Figure 3.1 illustrates the hurdles, these molecules had to cross before being prescribed as a drug.

There are two main approaches in Computer-Assisted Drug Design

◈ Structure based drug design (SBDD)

◈ Ligand based drug design (LBDD)

The structure here, implies the structure of the target (protein) and this approach is used when the identified protein for a particular disease has been structurally characterized by experimental means. Structure-aided drug design (SADD) requires crystals of protein complexed with ligands that are candidates for development as medicines. With X-ray crystallography these complexes are used to determine a three-dimensional model of the protein–ligand interaction at a molecular level. This information can then help-guide a rational design process in the development of the lead compound to become a medicinal candidate. Typically, an iterative process requiring structures with dozens of different compounds often of different structural classes is involved. This methodology proceeds to the selection of a small number of the best candidates, which are synthesized or purchased and tested for activity at the target. The results are then fed back into the CADD process.

Binding site is a cavity in the protein where the agonist/antagonist, interact or rather dock, to produce a change in the conformation of the protein due to which cell-to-cell signalling starts or stops as the case demands. One of the important means of identifying the binding site in a target is through co-crystallization, wherein the protein and the ligand are mixed in polar solvents and crystallized through special techniques such as 'hanging drop method'.

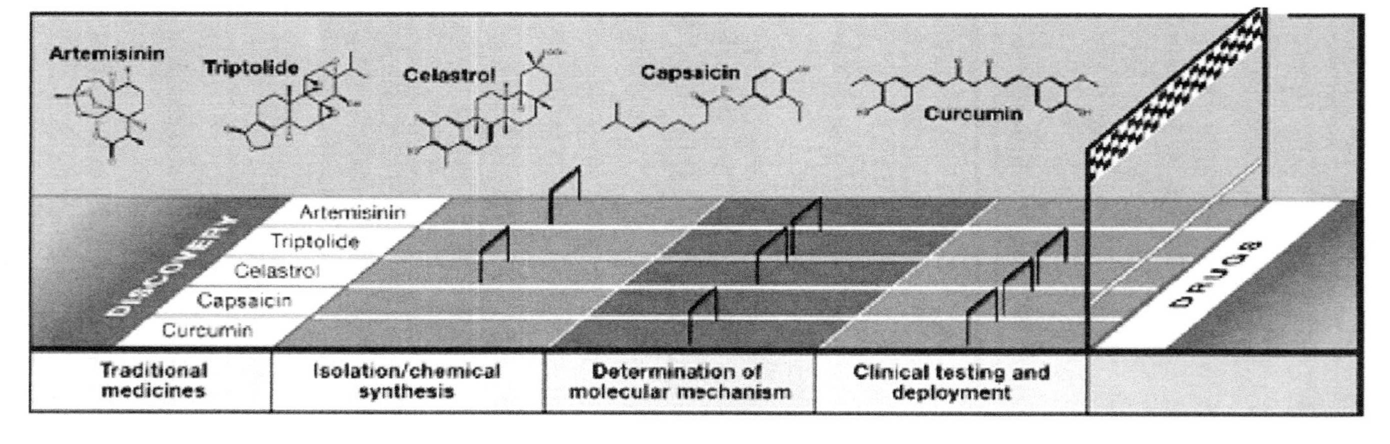

Figure 3.1 The route from traditional medicine to modern drug[8]

This may be successful by varying the pH, solvent composition, etc.

SBDD techniques include mainly **Docking** and **De novo design.**

LBDD approach is used when, knowledge about the target is not available or known. Quantitative Structure Activity Relationship Approach is used in such cases wherein chemical modification(s) of the key molecule followed by testing of the derivatives yield a structure–activity relationships (SARs), allowing a picture to be built up that is suggestive of the requirements of the receptor binding site.

Format Conversion

The two-dimensional or the three-dimensional structure of a molecule is represented by a file with a format depending on the need of the software used. For example, the output of an X-ray crystallographic study of a molecule would result in a **pdb** file which would contain the atomic coordinates of the atoms in the molecule and also a 'CIF' (Crystallographic Information File) which contains the information about the nature of data collection, size of crystals apart from the coordinates. Similarly, **sdf mol2, xyz** formats are also used for structure visualization.

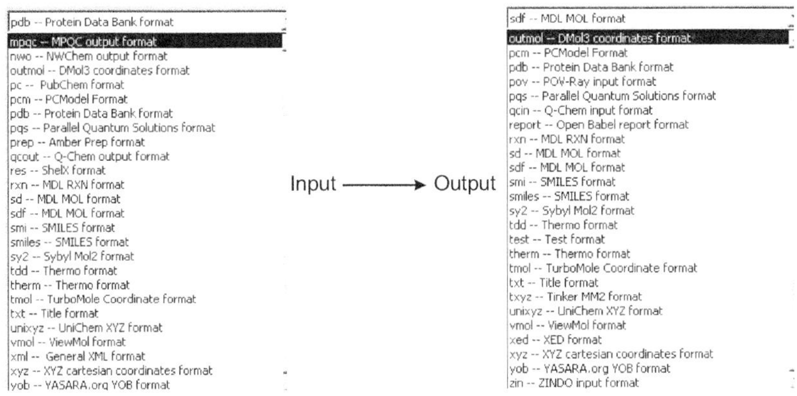

Figure 3.2   A snapshot of the input and output molecular formats in open babel

'Open Babel' (http://openbabel.sourceforge.net) is a very useful tool for inter-conversion of file formats, because certain softwares requires unique formats (like ∗.sdf) and if only the ∗.pdb file is available, it can be converted to the desired format (Figures 3.2 and 3.3).

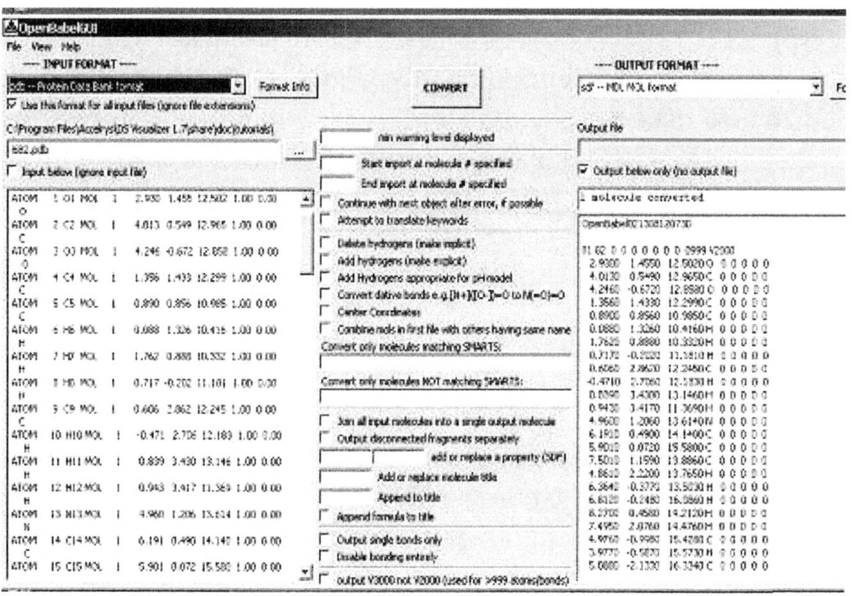

Figure 3.3   Conversion of a pdb file to sdf format

DOCKING

What is Docking?

In a majority of diseases, the progress of the disease, involves cell-to-cell communication in which proteins and neurotransmitters play an important role. These interactions result in a cascade of events, e.g. a catalytic reaction—like the cleavage of the substrate, or the stabilization of a transition state, or the blockage of the protein's active site due to the tight binding of an inhibitor, etc. Furthermore, these events lead to a sequence of steps that present the molecular basis of pharmacological effects.

The behaviour of small molecules in the binding pockets of target proteins can be described by molecular docking. If the type of interaction between the protein and the ligand is understood, it will form an efficient basis to design potent antagonists/drugs. This is the basic aim of docking.

Molecular docking can be defined as the prediction of the structure of receptor–ligand complexes, where the receptor is usually a protein and the ligand is either a small molecule or another protein. In other words, docking describes a process by which two molecules fit together in three-dimensional space. The crystalline structure of ligand bound to their target receptor is one of the most important sources to gain information about the basic mechanisms of interaction between the parts constituting the three-dimensional complex structure. In docking, the ligand and protein are predefined.

Types of Searching Methods in Docking

The ultimate aim of docking is to as far as possible, imitate the protein–ligand interactions perfectly, so that the results are dependable which can form the basis of further research and development. Based on the searching methodology involved, docking algorithms can be broadly classified into three categories.

1. *Searching the conformational space during docking*

Monte Carlo method, simulated annealing or genetic algorithm In **Monte Carlo** method, the system makes random moves and accepts or rejects each conformation based on Boltzmann probability. **Simulated annealing** is a generalization of the Monte Carlo method for examining the equations of state and frozen states of n-body systems. The basic concept of **genetic algorithm** is designed to simulate processes in natural system necessary for evolution, which is akin to the principles first laid down by Darwin. The newly generated solutions undergo selection, biased towards the fit among them. The algorithm maintains a selective pressure towards an optimal solution, with a randomized information exchange permitting exploration of the search space. A range of programs implements genetic algorithm for docking, including GOLD,[9,10] AUTO-DOCK,[11] and DARWIN.[12]

2. Searching the conformational space before docking

A conformational analysis is carried out first, and all relevant low-energy conformations are then rigidly placed in the binding site, where only the remaining six rotational and translational degrees of freedom of the rigid conformer must be considered. The programs **SLIDE** (<u>S</u>creening for <u>L</u>igands by <u>I</u>nduced-fit <u>D</u>ocking, <u>E</u>fficiently)[13] and **Fred** (<u>F</u>ast <u>R</u>igid <u>E</u>xhaustive <u>D</u>ocking)[14] use this docking methodology.

3. Incremental docking

In this method, the ligand is broken at rotating bonds creating fragments and these fragments are docked rigidly at various favourable positions in the binding site starting with a base fragment. This process is repeated until the entire ligand is assembled. Docking applications using incremental construction include FlexX,[15] Hammerhead,[16] HOOK,[17] and a component of DOCK4.0.[18]

Docking Methods

The complexity of computational docking increases in the following order:

i. *Rigid body docking*, where both the receptor and small molecule are treated as rigid.

ii. *Flexible ligand docking*, where the receptor is held rigid, but the ligand is treated as flexible

iii. *Flexible docking*, where both receptor and ligand flexibility is considered.

Most commonly used docking algorithms use the rigid receptor/ flexible ligand model.

It is well known that macromolecules often undergo conformational change, or induced fit, upon ligand binding in order to maximize energetically favourable interactions with the ligand or solvent.[19, 20] The driving force behind most induced fit mechanisms is hydrophobic interactions or hydrophobic collapse of the receptor around the bound ligand.[21] There are varying degrees of receptor flexibility. Conformational flexibility does not necessarily need to involve domain, tertiary, and/or secondary structure motions but may consist solely of subtle side-chain adjustments.

The Scoring Function

Detailed understanding of the general principles that govern the nature of the interactions (van der Waals, hydrogen bonding, electrostatic, etc.) between the ligands and their protein or nucleic acid targets, termed as the **'scoring function'** provides a conceptual framework for designing the desired potency and specificity. **Scoring** refers to evaluation and ranking of all configurations which are generated during a search process. The actual free energy is the preferred score.[22] This is not readily available for computation, so scoring functions must approximate binding free energies with sufficient accuracy (Figure 3.4).

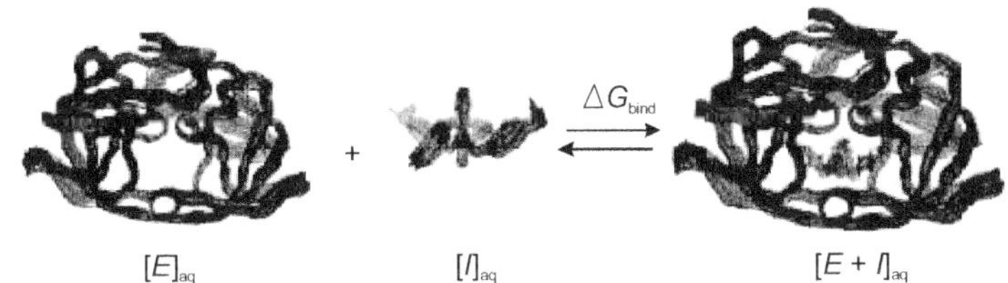

$[E]$—Enzyme

$[I]$—Inhibitor

$[E + I]$—Enzyme inhibitor complex all in aqueous environment.

Figure 3.4 A depiction of binding interaction between an enzyme and the inhibitor

The free energy of binding of the complex should be less than the sum of their individual free energy, i.e.,

$$\Delta G_{bind} \ [E + I] < \Delta G_{bind} \ [E] + \Delta G_{bind} \ [I]$$

Usually knowledge of interaction energies and forces flows into an assigned score. Scoring functions have a two-fold task.

1. They serve as an objective function to differentiate between diverse poses of a single ligand in the receptor-binding site.

2. After docking a compound database, they are required to estimate binding affinities of different receptor–ligand and to rank-order the compounds. Hydrophobic effects, van der Waal's and dispersion interactions, hydrogen bonding, steric, electrostatic interactions and solvation effects are among the factors that contribute to the ligand binding, which in turn are governed by kinetic and thermodynamic principles.

The binding affinity is estimated rather than calculated, accounting for the term 'scoring' function.

Types of scoring function

The available scoring functions may be categorized as

1. *Empirical-based* Empirical scoring functions sum enthalpic and entropic interactions with the relative weights of the terms based on a training set of protein–ligand complexes. The weights are assigned by regression methods that are used to fit the experimentally determined affinities. The interaction terms often include van der Waals, electrostatic interactions and hydrogen bonds.

Examples of empirical scoring functions include PLP,[23] ChemScore[24] and the FlexX[25] scoring function.

2. *Force-field based* Force field scoring functions are similar to empirical scoring functions in that they predict the binding free energy of a protein–ligand complex by adding up individual contributions from different types of interactions. However, they differ from empirical scoring functions in that the interaction terms are derived from physical–chemical phenomena as opposed to experimental affinities.

Examples of force field scoring functions in docking programs include the energy score in DOCK,[26] the score function used for single ligand docking in DOCKVISION[27] and that used in GOLD[28]

3. *Knowledge-based*   Knowledge-based scoring functions rely on statistical means to extract rules on preferred, and non-preferred, atom pair interactions from experimentally determined protein–ligand complexes. The rules are interpreted as pair-potentials that are subsequently used to score ligand binding poses.

Examples include the PMF score.[29]

4. *Consensus scoring*   This applies a number of score functions to the same docked pose identified by docking to eliminate false positives.[30,31]

Protein–ligand docking is so difficult due to tremendous complexity of the system; one must take into account hundreds of thousands of degrees of freedom in the two molecules, as well as the not-completely-known combination of energetic forces acting on them.[32] The dilemma becomes that while one needs to make simplifying assumptions to allow for reasonable computation time, these assumptions can also lead to unusable results.[33]

Factors affecting the docking score

- Orientation of residues

- Protonation states (ligand and protein)

- Tautomeric forms

- Protein flexibility

- Involvement of water

- Internal energy of the ligand

- De-solvation penalties

- Crystal packing of initial X-ray structure

Docking compounds from databases to targets of known structure can be utilized to discover or refine new leads. Structures from a database of small-molecule compounds are fit into the target structure using a docking program. The energies of the resulting complexes are evaluated and those that show the most promise can be experimentally tested as possible lead compounds.

Software for docking

There are many software programs available (both commercial and free) for docking. The robustness of a docking programme can be judged by its ability to reproduce the experimental result.

Free wares include

1. DOCK 6.1   http://dock.compbio.ucsf.edu/
2. AUTODOCK   http://autodock.scripps.edu/
3. HEX   http://www.csd.abdn.ac.uk/hex/ (protein– protein docking)
4. SLIDE   http://www.bch.msu.edu/~kuhn/ projects/slide/home.html

Commercial

1. GOLD   Cambridge Crystallographic Data Centre
2. FLEXX   Tripos, Inc
3. GLIDE   Schrodinger, Inc
4. ICM   Molsoft, Inc
5. LIGANDFIT   Accelrys, Inc
6. OPEN EYE

The raw materials for docking where to find protein and ligand structures?

Since docking is a structure-based drug design, the first question that has to be answered when planning a docking experiment is *"Is there an experimental structure for the protein that is to be used as a target during the docking?"*

To answer this question, it is necessary to go to the PDB database (http://www.pdb.org) and determine whether the three-dimensional structure of the corresponding target has been deposited or not. There are three important situations, one can expect in this case

1. The structure available is a co-crystallized one with a drug, which is the appropriate starting process for the docking.

2. The structure of the native protein alone is available, then the binding site has to be identified with appropriate software and then docking has to be progressed with.

   In both the above cases, the resolution cut-off of (~2.0 Å) and R-factor cut-off (0.250) should be taken into account, which reflects the accuracy in identifying the atomic coordinates.

3. If only the sequence of the target protein (and not the 3D structure) is available, then its similarity with protein sequences in database(s) is analysed. If a sequence with similarity ranging from 40% to 60% is identified, whose three-dimensional structure is available, then the 3D structure of the target protein is constructed through **homology modelling.**

CASE STUDY

Docking of Diclofenac to Cyclooxygenase-II

The aim of this exercise is to extract a drug that has been co-crystallized with a protein and re-dock it.

The protocol involved in docking is

1. Identify the disease/problem you have to work-on.

2. Browse in the protein databank (www.rcsb.org) for the target protein.

   For example, if one is planning to work on anti-inflammatory disease, Cyclooxygenase (COX) is one of the protein involved

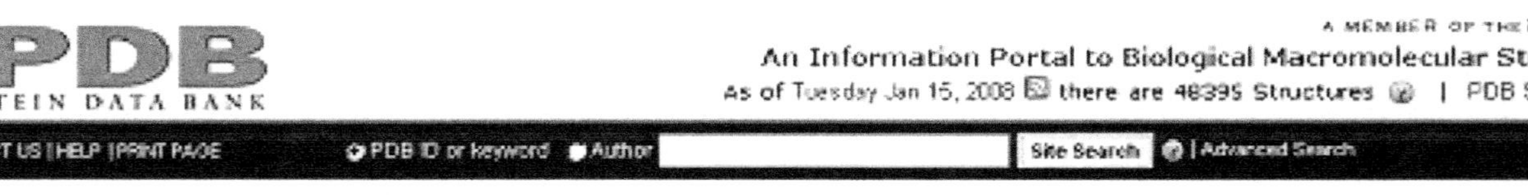

Figure 3.5 A snapshot of the search result using the term COX2

in the inflammation. Most of the drugs like ibrufen, diclofenac, etc., are targeted towards binding with this protein.

3. In the search box, type 'COX2' (the aim is to look for the structure of the protein bound/co-crystallized with the ligand).

4. This would result in the display shown in Figure 3.5.

5. There are three icons after the pdb ID 1pxx. Click the first icon to download the file.

6. Open the file with word pad.

7. The first few lines give the information regarding the protein (compound), its source and the publication information (Figure 3.6).

```
HEADER    OXIDOREDUCTASE                          07-JUL-03   1PXX
TITLE     CRYSTAL STRUCTURE OF DICLOFENAC BOUND TO THE CYCLOOXYGENASE
TITLE    2 ACTIVE SITE OF COX-2
COMPND    MOL_ID: 1;
COMPND   2 MOLECULE: PROSTAGLANDIN G/H SYNTHASE 2;
COMPND   3 CHAIN: A, B, C, D;
COMPND   4 SYNONYM: CYCLOOXYGENASE-2, COX-2, PROSTAGLANDIN-
COMPND   5 ENDOPEROXIDE SYNTHASE 2, PROSTAGLANDIN H2 SYNTHASE 2, PGH
COMPND   6 SYNTHASE 2, PGHS-2, PHS II, GLUCOCORTICOID-REGULATED
COMPND   7 INFLAMMATORY CYCLOOXYGENASE, GRIPGHS, TIS10 PROTEIN,
COMPND   8 MACROPHAGE ACTIVATION-ASSOCIATED MARKER PROTEIN P71/73,
COMPND   9 PES-2;
COMPND  10 EC: 1.14.99.1;
COMPND  11 ENGINEERED: YES
SOURCE    MOL_ID: 1;
SOURCE   2 ORGANISM_SCIENTIFIC: MUS MUSCULUS;
SOURCE   3 ORGANISM_COMMON: MOUSE;
SOURCE   4 GENE: PTGS2 OR COX2 OR COX-2 OR TIS10 OR PGHS-B;
SOURCE   5 EXPRESSION_SYSTEM: SPODOPTERA FRUGIPERDA;
SOURCE   6 EXPRESSION_SYSTEM_COMMON: FALL ARMYWORM;
SOURCE   7 EXPRESSION_SYSTEM_CELL_LINE: SF9;
SOURCE   8 EXPRESSION_SYSTEM_VECTOR_TYPE: BACULOVIRUS;
SOURCE   9 EXPRESSION_SYSTEM_PLASMID: PVL1393
KEYWDS    COX-2, CYCLOOXYGENASE, PROSTAGLANDIN, DICLOFENAC,
KEYWDS   2 ENDOPEROXIDE, OXIDOREDUCTASE
EXPDTA    X-RAY DIFFRACTION
AUTHOR    J.R.KIEFER,S.W.ROWLINSON,J.J.PRUSAKIEWICZ,J.L.PAWLITZ,
AUTHOR   2 K.R.KOZAK,A.S.KALGUTKAR,W.C.STALLINGS,L.J.MARNETT,
AUTHOR   3 R.G.KURUMBAIL
```

Figure 3.6 A snapshot of a pdb file-text format

8. In the pdb file, molecules other than the protein, are recognized by the identifier "HETNAM" including the water molecules found towards the end of the file.

In this pdb file the following four molecules are termed as the HETNAM in which diclofenac (DIF) is the drug molecule and the remaining are those accrued due to the buffers, etc., used under the crystallization conditions. Water molecules also are recognized as HETNAM and they need to be retained during the docking process.

HETNAM	NAG	N-ACETYL-D-GLUCOSAMINE
HETNAM	BOG	B-OCTYLGLUCOSIDE
HETNAM	HEM	PROTOPORPHYRIN IX CONTAINING FE
HETNAM	DIF	2-[2,6-DICHLOROPHENYL) AMINO] BENZENEACETIC ACID

9. The protein coordinates start from "ATOM'

ATOM	1	CA	ALA A	33	30.115	6.681–18.986	1.00	43.62	C
ATOM	2	C	ALA A	33	31.071	6.320–17.856	1.00	42.57	C
ATOM	3	O	ALA A	33	31.493	5.166–17.724	1.00	43.84	O
ATOM	4	CB	ALA A	33	30.900	7.160–20.219	1.00	43.09	C
ATOM	5	CA	ASN A	34	32.306	7.103–15.920	1.00	38.97	C
ATOM	6	C	ASN A	34	31.663	5.979–15.124	1.00	36.94	C
ATOM	7	O	ASN A	34	30.527	6.104–14.657	1.00	38.78	O
ATOM	8	CB	ASN A	34	32.392	8.383–15.072	1.00	42.24	C
ATOM	9	CG	ASN A	34	33.474	8.327–13.980	1.00	45.34	C

and terminates with "TER"

TER	17898	GLN	D3583

10. Following "TER" are the coordinates of the hetero atoms. Since we are interested only in diclofenac (drug), cut and copy the coordinates of the drug molecule alone into a word-pad file. Add the terms "REMARK4" as the header and "END" as the footer and save it as lig.pdb.

In this case, there are four molecules of DIF. It is enough if one molecule (say 701) alone is removed.

REMARK4

```
HETATM18390  C1   DIF   701   26.313  21.405   17.712  1.00  38.13  C
HETATM 18391 C2   DIF   701   26.065  22.511   16.868  1.00  39.49  C
HETATM18392  CL2  DIF   701   24.417  22.703   16.432  1.00  48.10  CL
HETATM18393  C3   DIF   701   27.128  23.265   16.312  1.00  40.58  C
HETATM18394  C4   DIF   701   28.456  22.836   16.649  1.00  40.32  C
HETATM18395  CL4  DIF   701   29.844  23.663   16.041  1.00  46.35  CL
HETATM18396  C5   DIF   701   28.692  21.725   17.480  1.00  38.88  C
HETATM18397  C6   DIF   701   27.618  21.016   18.014  1.00  40.33  C
HETATM18398  N1   DIF   701   27.058  24.431   15.437  1.00  35.21  N
HETATM18399  C7   DIF   701   26.417  25.622   13.353  1.00  28.89  C
HETATM18400  C8   DIF   701   26.533  24.442   14.127  1.00  30.87  C
HETATM18401  C9   DIF   701   26.213  23.207   13.552  1.00  31.18  C
HETATM18402  C10  DIF   701   25.752  23.134   12.235  1.00  30.32  C
HETATM18403  C11  DIF   701   25.600  24.293   11.481  1.00  29.18  C
HETATM18404  C12  DIF   701   25.928  25.525   12.039  1.00  29.13  C
HETATM18405  C13  DIF   701   26.884  26.964   13.814  1.00  28.04  C
HETATM18406  C14  DIF   701   28.360  27.023   13.980  1.00  28.76  C
HETATM18407  O1   DIF   701   29.107  27.680   13.260  1.00  28.91  O
HETATM18408  O2   DIF   701   28.809  26.265   14.998  1.00  29.05  O
END
```

11. Figure 3.7 shows the ball and stick representation (Rasmol) and the chemical structure of diclofenac obtained with the above coordinates as the input.

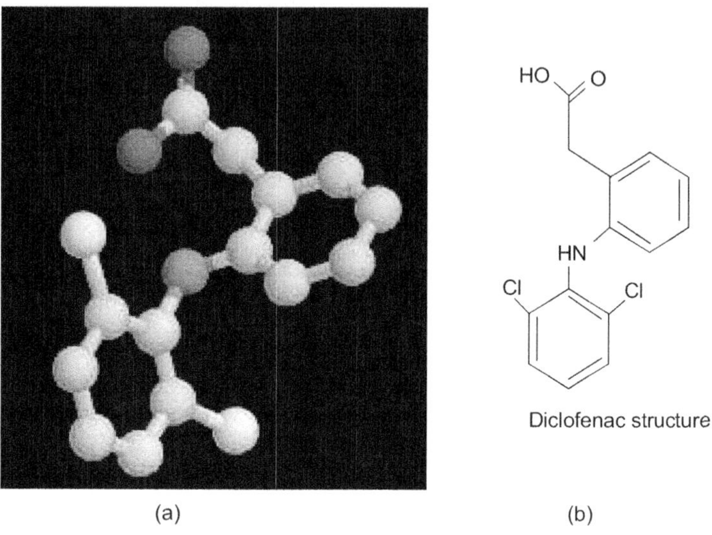

(a)   (b)

Figure 3.7 (a) Ball and stick representation of Rasmol (b) Diclofenac structure

What Have We Got Now?

We have now separated the protein and the ligand (from the binding site). The binding site is empty to go ahead with the docking process, wherein some or many selected molecules would be placed in it and ranked according t o their affinity in binding. The molecules for docking are selected from database basing on some criteria like similarity in structure or properties such as clogp, etc.

Docking Using AUTODOCK

AUTODOCK employs molecular dynamics to predict the docking of a flexible ligand to a binding site of a rigid protein, given the region of the protein containing the binding site and the substrate. For maximum predictive accuracy as well as reasonable computational time, it employs the AMBER force field in conjunction with free energy scoring functions and a large set of protein–ligand complexes with known protein–ligand constants.

AUTODOCK uses pre-calculated affinity maps for each atom type in the substrate molecule, usually C, N, O and H, plus an electrostatic map. These grids include energetic contributions from all the usual sources.

The energy (ΔG) contribution includes electrostatic interactions, hydrogen bond interactions, etc. and a typical equation relating them is given below.

$$\Delta G = \sum_{i,j} \left(\frac{A_{ij}}{r_{ij}^{12}} - \frac{B_{ij}}{r_{ij}^{6}} \right) + \sum_{i,j} E(t) \left(\frac{C_{ij}}{r_{ij}^{12}} - \frac{D_{ij}}{r_{ij}^{10}} + E_{hbond} \right)$$

$$+ \sum_{i,j} \frac{q_i q_j}{\varepsilon\left(r_{ij}\right) r_{ij}} + \Delta G_{tor} + \sum_{i_c,j} S_i V_j e^{\left(-r_{ij}^2 / 2\sigma^2\right)}$$

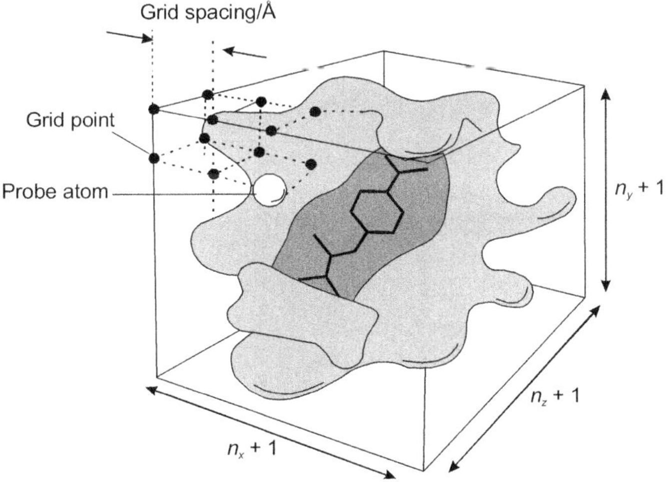

Each type of atom is placed at each individual grid point and the change in free energy is calculated. AUTODOCK can be downloaded from http://autodock.scripps.edu/along with Python Molecule Viewer (PMV).

1. Using the File pull down, read the molecules both protein.pdb and ligand.pdb onto the screen

2. Using the ligand pulldown use the input → Choose, this which will open up a screen from which choose lig.pdb and go to ouput submenu and save as lig.pdbq.

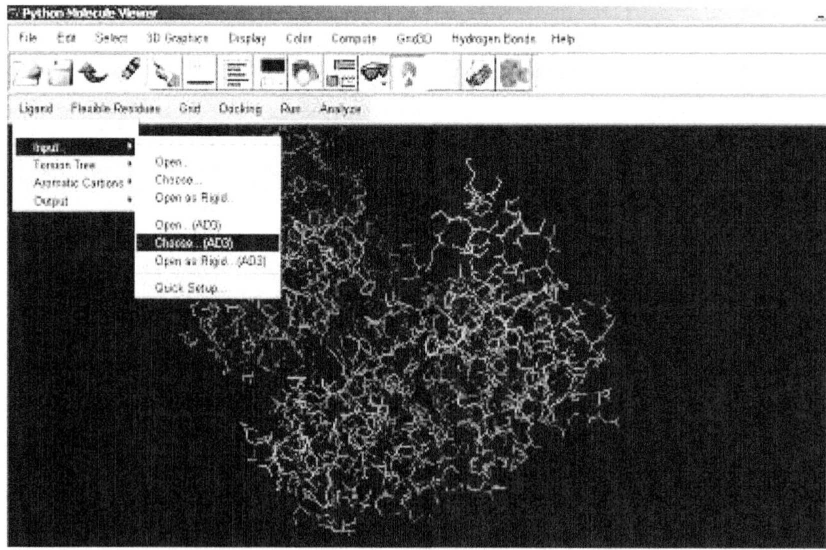

3. From the Grid pulldown, select Macromolecule →Choose.This will open up a box in which the lig.pdb and prot.pdb would be there. Choose prot.pdb and save as prot.pdbqs

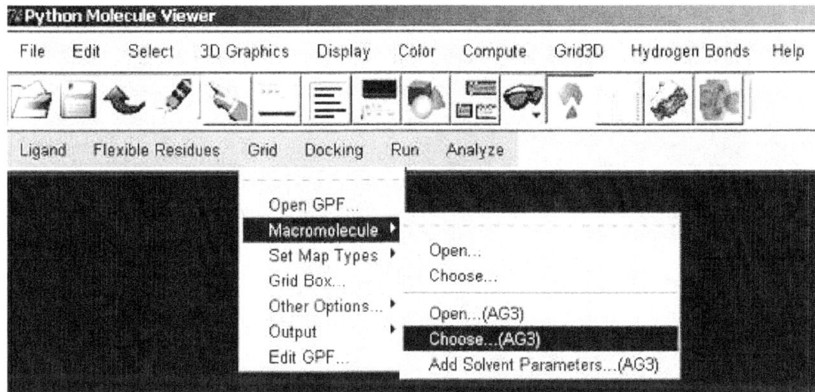

4. Set Map types →Choose ligand (choose the ligand from the box that opens). New Atom parameters have to be set if there are atoms that are not recognized by the program (for ex.Phosphorus)

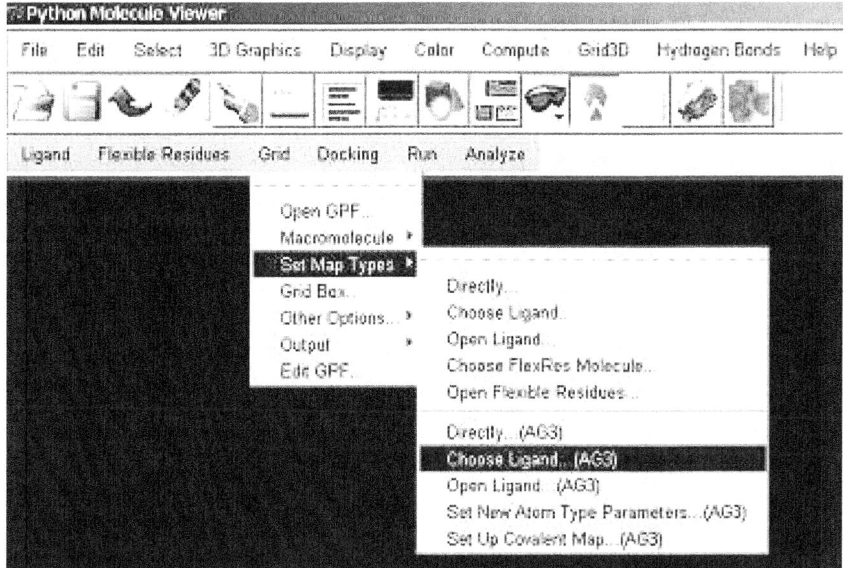

5. From the grid box pull down, select the dimensions of the grid box, in which the algorithm would search for the binding site. The box dimension can be set using, the three turn keys. After setting it, go to File → close saving current.

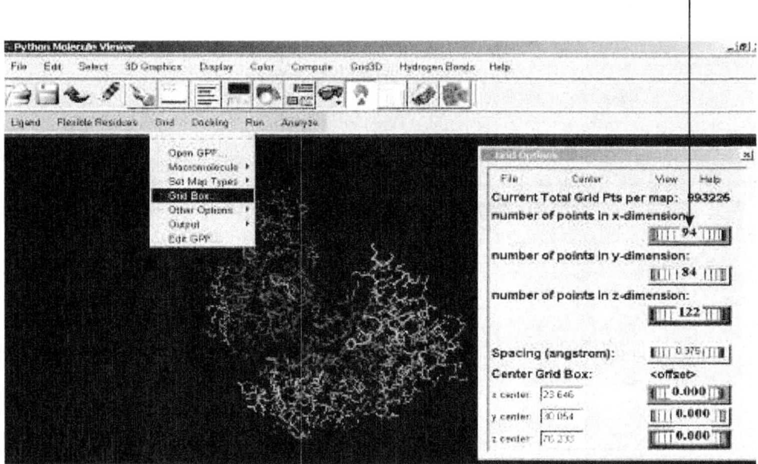

6. Go to other options, sub-menu and Accept the default values.

7. Then in the output sub-menu, choose → save as GPF.

8. On choosing edit GPF, the following screen will open. This Grid Parameter File contains information about the box dimension, the receptor molecule, types of atoms, etc.

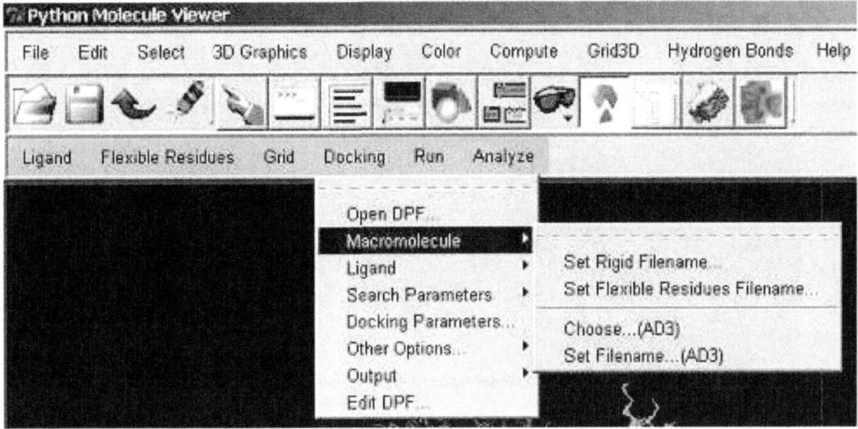

9. From the Docking pull down, select macromolecule → choose and choose the macromolecule.

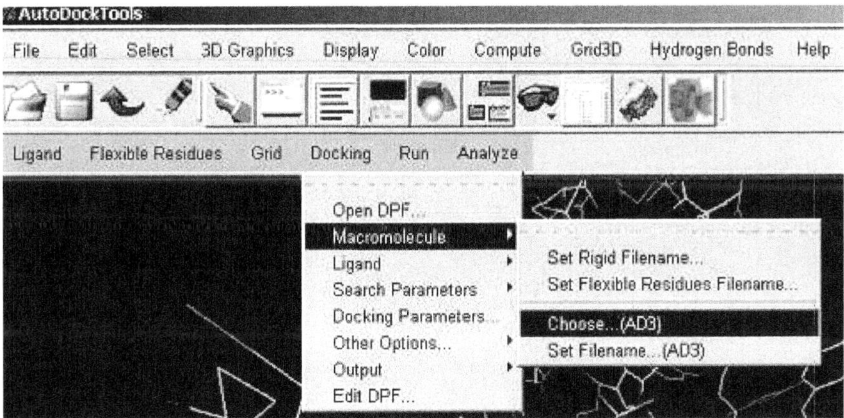

10. Go on to the third sub-menu Ligand → choose the ligand and also set the, ligand parameters.

11. In the search parameters, choose Genetic Algorithm.

12. Accept the docking parameters.

13. In the other options submenu, accept the AUTODOCK4 parameters.

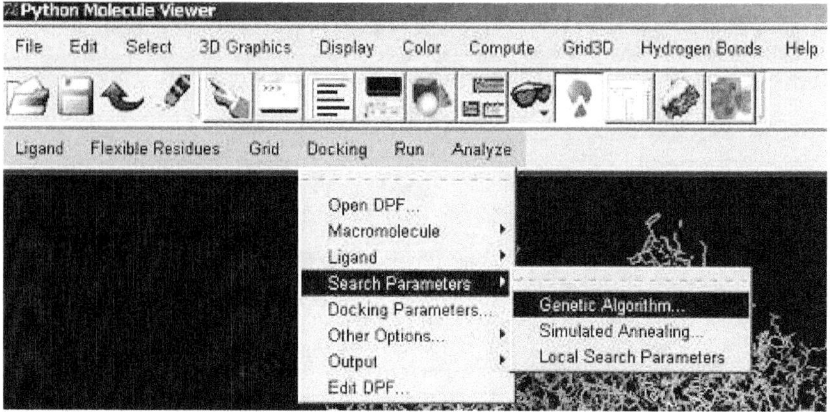

14. In the output menu, select Genetic algorithm, which will prompt you to save as ∗.dpf [docking parameter file].

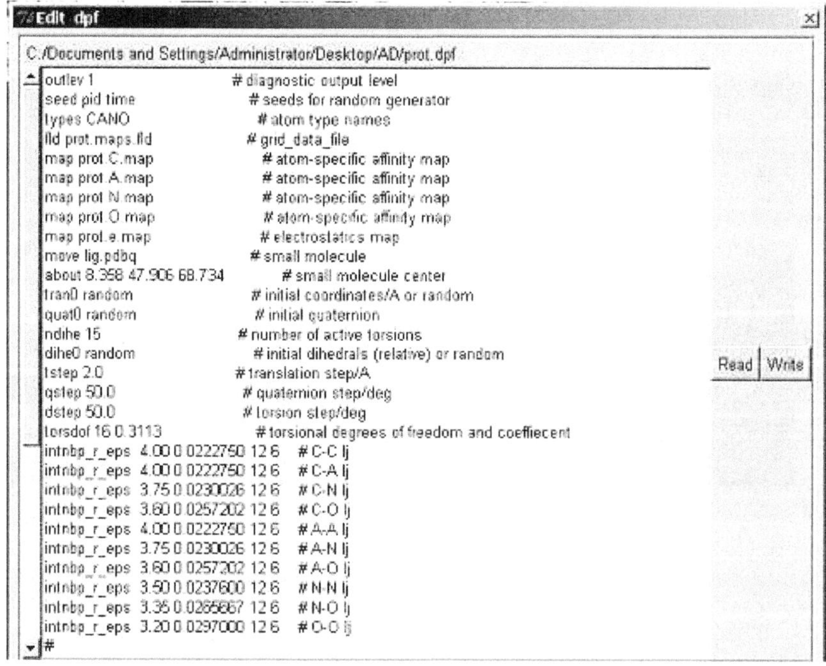

15. Select the Run menu → Run Autogrid

The program pathname should specify the location of autogrid3.exe

Parameter file name should be the *, gpf file

*.glg is the output file. Launch the run.

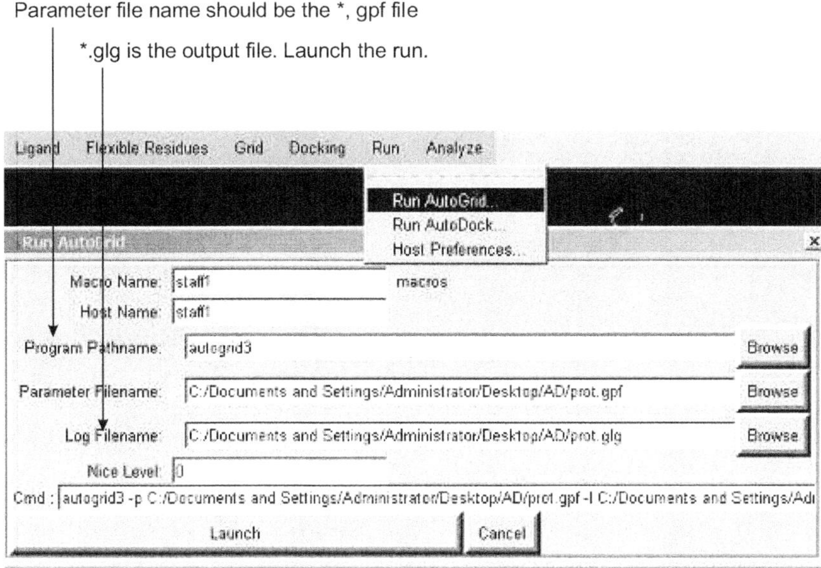

16. As the final step, Run AUTODOCK, for which $*$.dpf is the input file and $*$.dlg is the output file. This dlg (docking log file) contains the information regarding the binding energies of the different poses of the conformation along with their scoring.

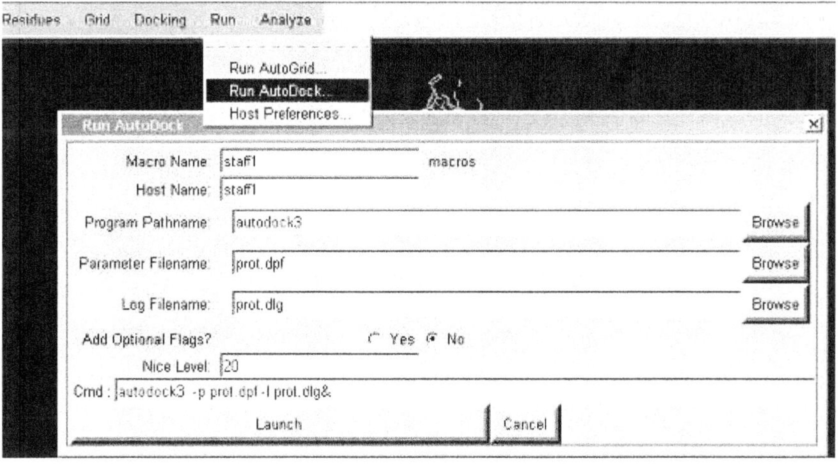

17. The final output file $*$.dlg would contain the binding energies of each model/conformation and are ranked. The

```
DOCKED: MODEL        1
DOCKED: USER      Run = 1
DOCKED: USER      DPF = prot.dpf
DOCKED: USER      Estimated Free Energy of Binding      =    -9.73 kcal/mol  [=(1)+(3)]
DOCKED: USER      Estimated Inhibition Constant, Ki     =    +7.42e-08       [Temperature = 298.15
K]
DOCKED: USER
DOCKED: USER      Final Docked Energy                   =   -12.47 kcal/mol  [=(1)+(2)]
DOCKED: USER
DOCKED: USER      (1) Final Intermolecular Energy       =   -12.22 kcal/mol
DOCKED: USER      (2) Final Internal Energy of Ligand   =    -0.26 kcal/mol
DOCKED: USER      (3) Torsional Free Energy             =    +2.49 kcal/mol
DOCKED: USER
DOCKED: USER      NEWDPF move lig.pdbq
DOCKED: USER      NEWDPF about 25.917299 17.174700 62.802898
DOCKED: USER      NEWDPF tran0 25.717572 17.199681 62.744870
DOCKED: USER      NEWDPF quat0 -0.198709 -0.979467 -0.034038 49677.952714
DOCKED: USER      NEWDPF ndihe 7
DOCKED: USER      NEWDPF dihe0 38139.16 -57295.72 53499.56 45365.26 -34929.56 -732.83 -5.62
DOCKED: USER                              x       y       z      vdW  Elec      q
DOCKED: USER
DOCKED: ATOM       1   C2  BCZ   801    24.460  16.576  63.135 -0.55 -0.13    +0.127
DOCKED: ATOM       2   C3  BCZ   801    25.526  17.671  62.840 -0.46 -0.06    +0.077
DOCKED: ATOM       3   C4  BCZ   801    25.525  18.562  64.132 -0.45 -0.08    +0.166
DOCKED: ATOM       4   O9  BCZ   801    26.468  18.000  65.089 +0.02 +0.14    -0.223
DOCKED: ATOM       5   C5  BCZ   801    24.085  18.406  64.730 -0.57 -0.06    +0.120
DOCKED: ATOM       6   C1  BCZ   801    23.337  17.434  63.783 -0.63 -0.04    +0.038
DOCKED: ATOM       7  N25  BCZ   801    24.044  15.973  61.868 -0.38 +0.09    -0.085
DOCKED: ATOM       8  C26  BCZ   801    23.563  14.780  61.848 -0.70 -0.89    +0.782
DOCKED: ATOM       9  N30  BCZ   801    23.297  13.998  62.942 -0.51 -0.07    +0.062
DOCKED: ATOM      10  N27  BCZ   801    23.228  14.231  60.632 -0.52 -0.08    +0.063
DOCKED: ATOM      11  C10  BCZ   801    26.962  17.154  62.473 -0.39 -0.05    +0.111
DOCKED: ATOM      12  N11  BCZ   801    26.890  16.195  61.363 -0.22 +0.11    -0.240
DOCKED: ATOM      13  C13  BCZ   801    27.233  14.893  61.561 -0.57 -0.05    +0.252
DOCKED: ATOM      14  C15  BCZ   801    27.079  13.932  60.396 -0.75 -0.02    +0.100
DOCKED: ATOM      15  O14  BCZ   801    27.662  14.422  62.604 -0.09 -0.04    -0.271
DOCKED: ATOM      16  C24  BCZ   801    27.938  18.346  62.113 -0.42 -0.01    +0.023
DOCKED: ATOM      17  C37  BCZ   801    27.615  18.966  60.720 -0.61 -0.00    +0.002
DOCKED: ATOM      18  C38  BCZ   801    27.742  20.501  60.591 -0.75 -0.00    +0.000
DOCKED: ATOM      19  C36  BCZ   801    29.409  17.818  62.212 -0.44 -0.00    +0.002
DOCKED: ATOM      20  C39  BCZ   801    30.255  18.471  63.328 -0.40 +0.00    +0.000
DOCKED: ATOM      21   C6  BCZ   801    23.357  19.699  64.931 -0.63 +0.04    +0.187
DOCKED: ATOM      22   O7  BCZ   801    22.439  19.734  65.897 -0.11 -0.42    -0.647
DOCKED: ATOM      23   O8  BCZ   801    22.604  20.729  64.305 -0.15 -0.31    -0.647
DOCKED: TER
```

model with the least binding energy (ies) is/are taken as the best model and selected for hydrogen bonding interaction calculations.

Importance of Hydrogen Bonds in Protein–Ligand (Drug) Interactions

Hydrogen-bonding interactions play a pivotal role in the action of both agonists and antagonists. Both, through hydrogen bonds, invoke a conformational change in the protein with results in the desired action. The bonding energies of hydrogen bonds (4–40 kcal/mol) are lower than those of covalent bonds. Accordingly hydrogen bonds like O—H···O, N—H···O, O—H···N and N—H···N (Figure 3.8) may be considered to be strong while interactions like C—H···O, C—H···N, O—H··· π , N—H··· π and C—H··· π are taken as weak. Hydrogen atoms bound to nitrogen and oxygen atoms form hydrogen bonds with lone electron pairs on other oxygen and nitrogen atoms (Figure 3.8). These 'classical' hydrogen bonds have been held responsible for the formation of secondary structural elements such as α helices and β sheets and, along with van der Waals and hydrophobic forces, they constitute one of the main pillars of overall protein stability and a principal determinant of protein conformation. But apart from this, the weaker hydrogen bonds C−H...O and C−H... π have also been reported.[34]

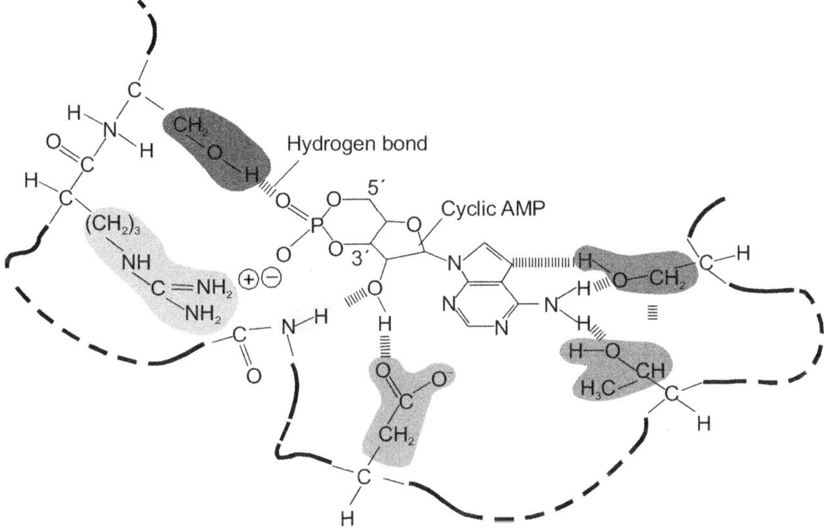

Figure 3.8 A depiction of hydrogen-bonding interactions and ionic interaction between the ligand (cyclic AMP) and the protein

Hydrogen bonds play a crucial role in preserving the three-dimensional shape of a protein or DNA. The double helix of DNA, RNA structures, peptide and protein secondary structures, like α-helices, β-sheets, β- and γ-loops, and the tertiary structures of proteins are formed by hydrogen bonds. In drug design, it is most important to realize that ligands are recognized by properties, not by their molecular structure. Analogues which exert similar interactions with the functional groups of the binding site have similar affinities, even if their chemical structures are quite different.

The hydrogen bond itself gives rise to its own classification on the basis of energetic criteria

i. Weak hydrogen bond is characterized by an intermolecular energy of the order of 20 kJ/ml or less. The weak hydrogen bond may be defined as an interaction D---H•••A (D–Donor, A– Acceptor), wherein a hydrogen atom forms a bond between two structural moiety X and A, of which one or even both are only of moderate or low electronegativity.[35]

ii. Medium hydrogen bond, for which the complexation energy is ca 20–50 kJ/mol, and

iii. Strong hydrogen bond with an interaction energy 80–150 kJ/mol.

Difference in the structures need not indicate difference in hydrogen bonding pattern. For example, scytalone dehydratase is an enzyme that converts scytalone to 1, 3, 8-trihydroxynaphthalene-on the melanin biosynthesis pathway. Inhibitors for this enzyme, salicylamides and quinazolines bind to scytalone dehydratase in a comparable manner. The hydrogen bonds between the interacting groups are identical, despite the differences in the chemical structures[36] (Figure 3.9).

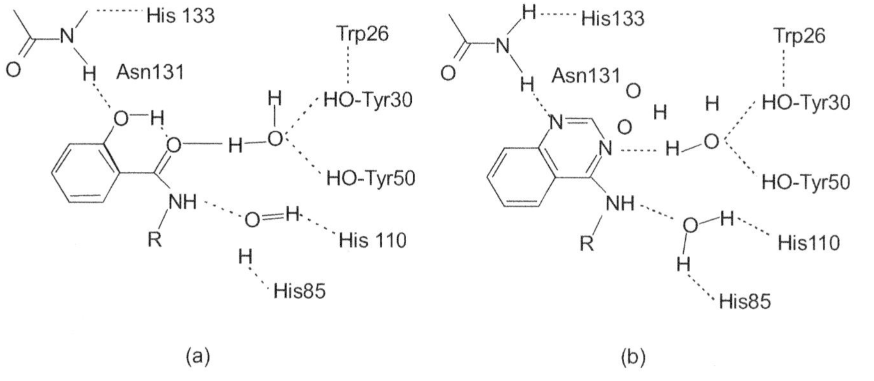

Figure 3.9 Similarity in the hydrogen bonding pattern of salicylamide derivative and quinazoline derivative

Hydrogen bonding may be represented as D---H•••A. In general, D has been found to be more electronegative than H and the regions of high electron density could be a lone pair of electrons in an acceptor atom, A or π electron cloud in a molecule.

HBAT—A Tool for Analysing Hydrogen Bond Interactions[37]

The program HBAT is a tool to automate the analysis of potential hydrogen bonds and similar type of weak interactions like halogen bonds and non-canonical interactions in macromolecular structures, available in Brookhaven Protein Database (PDB) file format. HBAT supports post-docking interaction analysis between PDB files for any target/receptor (in PDB files) and docked ligands/poses (in SDF). This tool can be implemented in active site interaction analysis, structure-based drug design and molecular dynamics simulations.

CASE STUDY

1. Download the HBAT program from http://202.41.85.161/ ~grd/HBAT.html and install it (runs on windows platform).

2. Download the pdb file **1L7F** from the brookhaven database. This is the co-crystal structure.

3. Since this program requires hydrogens to be fixed on the proteins and ligands, fix hydrogens using AccelrysDSvisualizer [can be freely downloaded from www.accelrys.com] or Python Molecular Viewer (PMV) [http://www.scripps.edu/~sanner/python/pmv/ index.html] and save the pdb in the same name.

4. Using the File option, open the pdb file which would result in the screen shown below.

5. Set the parameters, to either the default values or the user-defined values. Default values would result in listing of all the hydrogen bonds in the protein. By specifying the atom number only the hydrogen bonds around the specified

atoms (ligand), with a cut-off, say 10 Å) would be calculated. In this case specify the atom number as 6272.

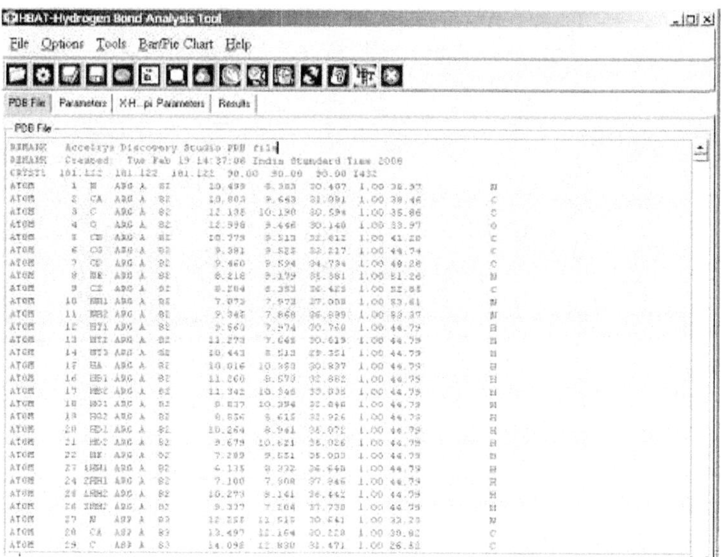

6. The X-H...pi parameters can be left as default unless necessary. This would calculate the weaker hydrogen bonds such as C—H...O, C—H... π, etc.

7. Choose 'local interaction analysis' and 'cooperative graph' from the options menu and then from the tools menu click on 'start calculations'. This would result in the list of hydrogen bonds as given below, which would include the donor–acceptor, the distance between them, type of hydrogen bonds, the angle between them, etc.

```
 HB List | Frequency | Distribution | Furcations | Halogen Bonds | X-H...pi

 Intermolecular H-Bond List
 C-H...O (S NP S CH) ILE( CD1 -  427)   H GLU( OE1 -  425)      1.090 2.552 3.473 141.6
 C-H...O (S NP W **) ILE( CD1 -  427)   H HOH( O   -   46)      1.090 2.716 3.508 129.0
 C-H...O (L ** S CH) BC2( C2  -  801)   H ASP( OD1 -  151)      1.089 2.558 3.333 127.2
 C-H...O (L ** S CH) BC2( C2  -  801)   H ASP( OD2 -  151)      1.089 2.842 3.720 137.6
 N-H...O (L ** W **) BC2( N25 -  801)   H HOH( O   -   35)      1.070 2.148 2.897 124.9
 N-H...O (L ** S CH) BC2( N30 -  801)   H GLU( OE2 -  119)      1.069 2.585 3.441 136.5
 N-H...O (L ** S CH) BC2( N30 -  801)   H ASP( OD1 -  151)      1.069 2.522 3.123 114.6
 N-H...O (L ** W **) BC2( N30 -  801)   H HOH( O   -   19)      1.069 2.461 3.439 151.5
 N-H...O (L ** B CH) BC2( N30 -  801)   H ASP( O   -  151)      1.069 2.949 3.169 91.78
 N-H...O (L ** S CH) BC2( N30 -  801)   H ASP( OD1 -  151)      1.069 2.790 3.123 97.86
 N-H...O (L ** S CH) BC2( N27 -  801)   H GLU( OE1 -  227)      1.070 2.445 3.030 113.1
 N-H...O (L ** S CH) BC2( N27 -  801)   H GLU( OE1 -  227)      1.070 2.705 3.030 97.13
 C-H...O (L ** S CH) BC2( C3  -  801)   H GLU( OE2 -  277)      1.090 2.798 3.885 174.4
 C-H...O (L ** S  P) BC2( C3  -  801)   H TYR( OH  -  406)      1.090 2.982 3.657 120.3
 C-H...O (L ** W **) BC2( C3  -  801)   H HOH( O   -   35)      1.090 2.854 3.616 126.9
 C-H...O (L ** S CH) BC2( C37 -  801)   H GLU( OE2 -  277)      1.089 2.597 3.509 140.6
 C-H...N (L ** S CH) BC2( C38 -  801)   H ARG( NH1 -  292)      1.090 2.805 3.806 152.5
 C-H...N (L ** S CH) BC2( C39 -  801)   H ARG( NH2 -  152)      1.089 2.995 3.949 146.4
 O-H...N (L ** S CH) BC2( O14 -  801)   H ARG( NH2 -  152)      1.050 2.268 2.852 113.2
 O-H...O (L ** S CH) BC2( O9  -  801)   H ASP( OD2 -  151)      1.050 2.046 3.060 161.0
 C-H...O (L ** W **) BC2( C5  -  801)   H HOH( O   -  209)      1.090 2.931 3.944 154.6
 C-H...N (L ** S CH) BC2( C6  -  801)   H ARG( NH2 -  118)      1.089 2.863 3.654 129.4
 O-H...N (L ** S CH) BC2( O8  -  801)   H ARG( NH1 -  118)      1.049 2.385 3.113 125.4
 O-H...O (L ** S  P) BC2( O8  -  801)   H TYR( OH  -  406)      1.049 2.795 3.309 110.2
 O-H...O (L ** W **) BC2( O7  -  801)   H HOH( O   -  301)      1.050 1.896 2.799 141.8
 C-H...O (L ** S  P) BC2( C1  -  801)   H TYR( OH  -  406)      1.090 2.350 3.248 138.3
 C-H...O (L ** S CH) BC2( C1  -  801)   H GLU( OE2 -  119)      1.090 2.559 3.300 124.3
 C-H...O (L ** S CH) BC2( C1  -  801)   H ASP( OD1 -  151)      1.090 2.653 3.320 118.8
 O-H...O (W ** B  P) HOH( O   -   16)   H THR( O   -  225)      1.050 2.667 3.693 165.5
 O-H   O (W ** W **) HOH( O   -   16)   H HOH( O   -   32)      1.050 2.025 2.696 118.9
```

8. From the tools menu, select "Prepare co-operative graph" which would result in a 'Dotty graph' as represented below. This would clearly indicate as to how the atoms are linked through hydrogen bonds.

9. The results can be saved as $*$.hbat and contains information about the hydrogen bonds and also the bifurcated interactions.

The treatment of conserved water molecules and small ions is an important outstanding problem in atomistic docking models.[38] Removal of these molecules increases the size of the cavity, leading to potentially incorrect results, while their explicit treatment would increase the size of the search space drastically and place high demands on the accuracy of the scoring function.

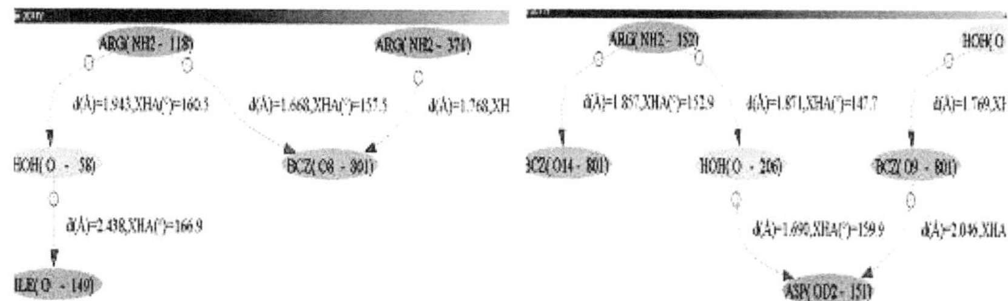

A dramatic illustration of the influence of crystal water is 3cla,[39] where the ligand is almost entirely embedded in water. Various water molecules are located in the neighbourhood of the ligand to stabilize the binding mode. Although some scoring functions (e.g. Glide, AUTODOCK) implicitly account for solvation effects, this approach remains fruitless here, because the water molecules serve as bridges that mediate the ligand–target interaction.

VIRTUAL SCREENING

The aim of drug discovery is to identify molecules that will lead to clinically safe and effective new medicines. High-throughput screening (HTS) is traditionally recognized as automated screening of chemical libraries (10^3 to 10^6 molecules) and data capture. In HTS, chemical libraries can be directed against specific target classes (receptors, kinases), full-file compound collections, or selective chemical series.

Virtual screening is the process by which computational tools are used to reduce the number of actual compounds screened.[40] It can be considered as an *in silico* counterpart of HTS. Pharmaceutical companies and chemical vendors have very large databases of chemical compounds (millions) available for screening. Their goal is to reduce the number needed to screen while

⬥ Increasing the probability of biological activity

⬥ Increasing the probability of oral absorption

⬥ Decreasing the probability of toxicity

⬥ Decreasing the number of false positives

Massive throughput in synthesis and screening yields, on an average, hundreds of thousands of novel compounds that are synthesized, then screened for various properties, ranging from biological activity, solubility , metabolic stability and cytotoxicity. The structure-based virtual screening begins with the identification of a potential ligand-binding site on the target molecule. Ideally the target site is a pocket or protuberance having a variety of probable hydrogen bond donors and acceptors, hydrophobic characteristics, and with molecular adherence surfaces. The ligand-binding site can be the active site as in an enzyme an assembly site with another macromolecule or a communication site, which is necessary in the mechanism of the molecule. Determining the structure of a target protein by NMR, X-ray crystallography or homology modelling befalls as a major and initializing stair in structure-based virtual screening. Numerous X-ray crystallographic and NMR studies in determining the experimental structures of ligands bound to the enzymes serve as a major source of ideas for analog design, in turn useful for the docking studies.

While wet-lab screening can conduct a high throughput screening of 10,000 –100,000 compounds/day, virtual screening can accomplish the same for 5–20 minutes per compound per CPU. Also the hit rate is 1 in 100,000 (non-focused libraries) in the case of the former and approximately 10,000 (64 CPU cluster, structure-based docking/screening) compounds in the latter.

The principle here is to dock all the ligands present in a database into the binding pocket of the selected target and evaluate the fit between the molecules. The quality of the fit is then used to rank the small molecules. To perform SB-VLS computations, many tools are usually needed: protein structure prediction, structural analysis, compound collections, *in silico* ADME/tox prediction, definition of ligand binding site (selection of targets) and *in silico* docking and scoring methods.

Since any docking tool needs to combine a docking engine with a fast-scoring function, the success of virtual screening relies on three possible issues:

1. the capability of a docking algorithm to reproduce the X-ray pose of selected small-molecular-weight ligands[41-43]

2. the propensity of fast-scoring functions to predict binding free energies from the best-scored pose[44]

3. the discrimination of known binders from randomly chosen molecules in virtual screening experiments

PITFALLS IN DOCKING

The quality of the receptor structure—in terms of resolution— employed plays a central role in determining the success of docking calculations.[41,42] In general, the higher the resolution of the employed crystal structure, the better the observed docking results.

Structure refinement protocols of protein–ligand complexes,[43] the pH dependence of ligand binding modes, uncertainties in locating the ligands ('mistaken identity') in the co-crystal structures as well as the subjective nature of deriving good quality protein models are deciding factors in a fruitful docking process. Even at high resolution, the difficulties in defining ligand atomic positions unambiguously can be attributed to the disparity between the high-quality dictionaries of bond lengths, bond angles and torsions available for protein and nucleic acid structure refinement and those available for small organic molecules. Regardless of the ambiguities, success has been reported for numerous high throughput docking studies using X-ray receptor structures. Recent examples of this type of study include: thymidylate synthase,[45] phosphoribosyl transferase,[46] farnesyl transferase,[47] FKBP12.[48]

DEPICTING THE PROTEIN–LIGAND INTERACTION

Ligplot[49] is a tool for plotting the protein–ligand interaction including hydrogen bonds, ionic interactions, etc. Ligplot can be downloaded from http://www.biochem.ucl.ac.uk/bsm/ligplot/ligplot.html. It is free for academic users. Given below is the model of rolipram bound at the active site of PDE4[50] generated using Ligplot (Figure 3.10). Hydrogen bonds are presented as dashed lines, and the interatomic distances are shown in Angstroms. The residues that form van der Waals contacts with docked rolipram are depicted as labelled arcs with radial spokes that point toward the ligand atoms with which they interact.

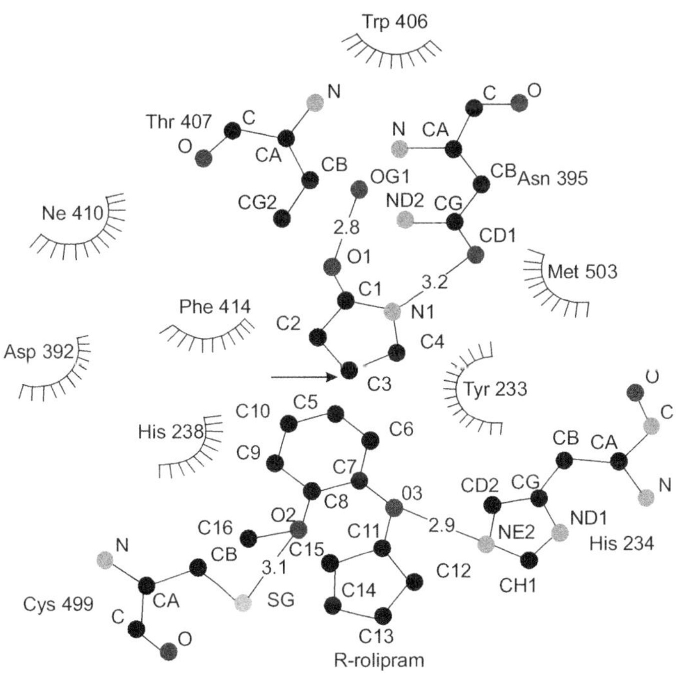

Figure 3.10   Rolipram bound at the active site of PDE4

PHARMACOPHORIC APPROACH

Pharmacophore, which is the three-dimensional arrangement of essential atomic features that enable a molecule to exert a particular biological effect, is a very useful model for achieving success in database searching and virtual screening.[51] In this model the rest of the molecule acts as a scaffold to hold the groups in the right place. The pharmacophore can be generated by visual inspection or by computational techniques. Typically, the derived pharmacophores consist, generally, of 3–5 features such as charged centres, hydrogen bond donors and acceptors, hydrophobic centres, aromatic ring centres, etc. The distances between them, angles and other geometric measures are sometimes used. By examining the chemical properties and the possible shapes of these ligands, a set of features are identified (Figure 3.11).

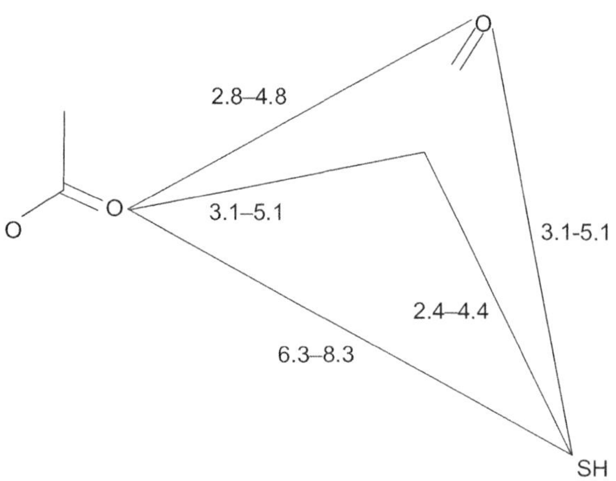

Figure 3.11   A pharmacophore

The features of the pharmacophore interact with features of the receptor, while the rest of the ligand acts as a scaffold. Once a pharmacophore has been isolated, it can be used to further improve the activity of a pharmaceutical drug. If the three-dimensional structure of the receptor is known, pharmacophore is a complementary tool to standard techniques, such as docking. However, frequently the structure of the receptor protein is unknown and only a set of ligands together with their measured binding affinities towards the receptor is available. In such a case, a pharmacophore-based strategy is one of the few applicable tools.

There are two different ways to deduce a pharmacophore(s) (Figure 3.12): direct methods and indirect methods. Direct methods use both ligand and receptor information, whereas indirect methods use only a collection of ligands that have been experimentally observed to interact with a given receptor. Frequently, the exact structure of the receptor is unknown. In such a situation, only indirect methods are applicable. However, direct methods are becoming extremely important with the rapidly increasing number of known protein structures, which is the outcome of the structural genomics project.

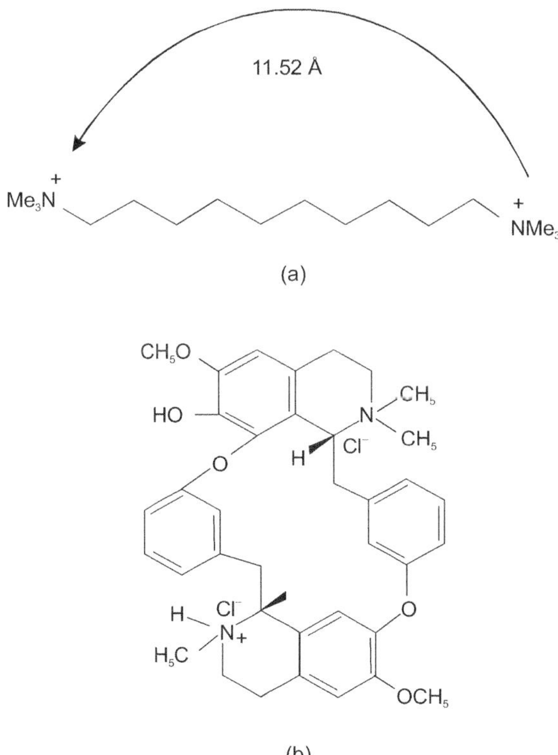

Figure 3.12 (a) A pharmacophore-based analogue of neuromuscular blocking drug decamethonium (b) Tubocurarine chloride

DE NOVO DESIGN

If one fails to find a drug molecule having the required interacting groups by database searching, the alternative may be to construct a ligand having active groups placed in such a way that interaction with the protein at the identified interaction sites is possible. This ligand construction process is called de novo ligand design. It involves the design of active compounds including novel structures that do not exist in known compound databases. In most cases, the design is based on the target receptor's 3-dimensional (3D) structures including X-ray structures, homology models, and pharmacophore maps. Compared with virtual screening by docking of known molecules in compound databases (3D database search), de novo ligand design can generate novel active structures fitted to the active site of the target protein and open the door to utilize the whole chemistry space. A large number of de novo design programs are available. These can be divided into three main categories (Figure 3.13):

1. Those that connect molecular fragments placed at the interaction sites to obtain a ligand in a process known as **Linking.**

2. Those that start from one fragment and connect fragments sequentially to it using random connection methods which is termed as **Growing**.

3. The last category includes the genetic algorithm methods. Most of the random connection methods start from an initial "pool" of fragments and construct ligands by making and breaking connections between the fragments.

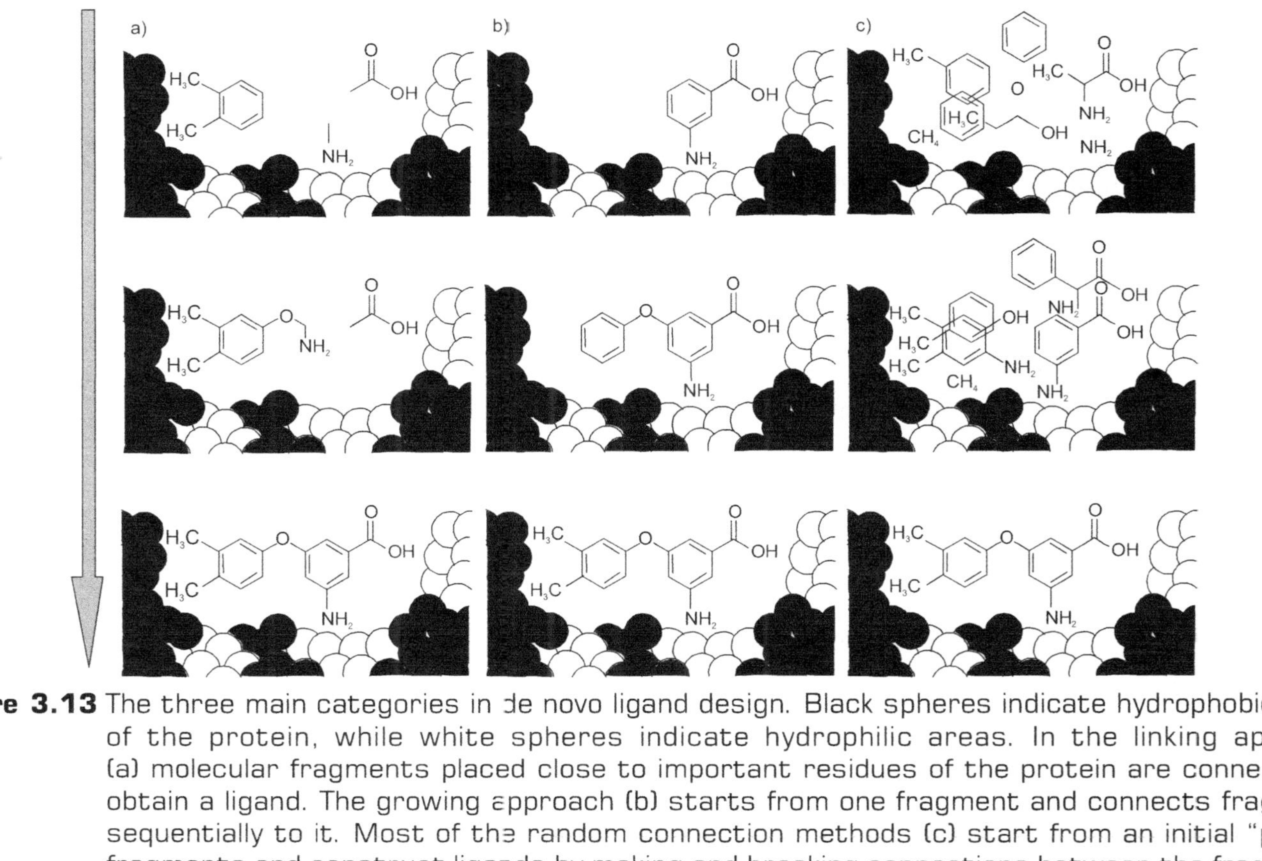

Figure 3.13 The three main categories in de novo ligand design. Black spheres indicate hydrophobic areas of the protein, while white spheres indicate hydrophilic areas. In the linking approach (a) molecular fragments placed close to important residues of the protein are connected to obtain a ligand. The growing approach (b) starts from one fragment and connects fragments sequentially to it. Most of the random connection methods (c) start from an initial "pool" of fragments and construct ligands by making and breaking connections between the fragments.

Table 3.1 A list of de novo software and their web addresses

Method	Type[b]	URL
BUILDER	L	http://thalassa.ca.sandia.gov/-dcroe/
CAVEAT	L	http://www.cchem.berkeley.edu/~pabgrp/data/caveat.html
HOOK	L	http://www.accelrys.com/quanta/mcss_hook.html
LUDI	L	http://www.accelrys.com/insight/ludi.html
PRO_SELECT	L	http://www.protherics.com/wtech_camdt.html
SKELGEN	L	http://www.denovopharma.com
SmoG	G	http://www-shakh.harvard.edu/-smog/
CombiSMoG	G	http://www.concurrentpharma.com
SPLICE	L	http://www.tripos.com
SPROUT	G	http://www.simbiosys.ca/sprout/
Lig Builder	L + G	http://mdl.ipc.pku.edu.cn/drug_design/work/ligbuilder.html
Leapfrog	G	http://www.tripos.com
DycoBlock	L	yyshi@iris.bio.ustc.edu.can
ADAPT	R	http://mako.cgl.ucsf.edu/~spegg/
LEA	R	douguetl@caramail.com

[b]L—linking approach, G—growing approach

R—random connection approach

"De novo" means "novel". Database searching and docking (virtual screening) would result in only known molecules but with different activity.

De novo drug design is an iterative process in which the three-dimensional structure of the receptor is used to design newer molecules. It involves structure determination of the lead target complexes and the design of lead modifications using molecular modelling tools. It can also be used to design new chemical classes of compounds that present similar substituents to the target using a template or scaffold, which is chemically distinct from previously characterized leads. Ligand-based de novo design is not provided with interaction sites as primary target constraints. The majority of ligand-based methods operates on the topological molecular graphs, and features an evolutionary algorithm for optimization.

Molecular de novo design produces novel molecular structures with desired pharmacological properties from scratch. In this approach, a medicinal chemist—and, equally, the de novo molecule-design software—is confronted with a virtually infinite search space. The number of chemically feasible, drug-like molecules has been estimated to be in the order of 10^{60}–10^{100}, from which the most promising candidates have to be selected. In de novo design there is no direct relation to known leads.

The steps involved in the de novo design process are:

1. Identification of the binding site in the receptor

2. Evolving a pharmacophore from the binding site residues

3. Construction/identification (through database searching) of the ligands fitting the pharmacophore

4. The scoring of the structures

Structure-based lead generation using these programs has some merit compared with 3D database searches of known compounds. First, as already mentioned, de novo design program can construct structures including novel skeletons. Moreover, methods that dock whole molecules at a time like a 3D database search have one or more of the following problems.

1. The number of considered conformations for each ligand is not sufficient.

2. Some functional groups work as ionic forms under certain physiological conditions (pH 7.4), whereas docking programs often use only neutral forms.

3. Even the largest collections will be biased toward particular types of compound. The bias may result from, among other things, the historic interest of a company in a given

therapeutic area or the academic interests of those depositing crystal structures.

In the case of de novo design, considering conformations of functional groups would not be a critical problem because de novo design programs generate functional groups in appropriate forms and conformations to interact with the protein surface. However, in most cases, de novo design programs also have the following critical defects.

1. de novo design programs do not consider synthetic feasibility when constructing structures

2. the prediction of binding affinities for the designed structures is not so accurate. For these reasons, de novo-designed structures have been synthetically problematic with insufficient and absent potency.

Therefore, it has been practically difficult to obtain novel lead compounds that are appropriate for medicinal chemists in actual drug discovery projects from de novo-designed structures.

THE PRINCIPLE OF DE NOVO DESIGN

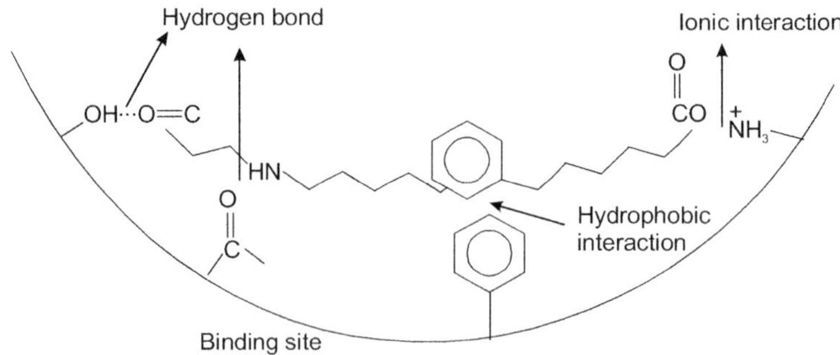

Figure 3.14 A depiction of functional group residues in the binding site and the possible complementary groups that can interact with it

The binding site contains amino acids carrying varying substituent groups such as hydroxyl, carbonyl, protonated amine, etc. (Figure 3.14). From this information the complimentary functional group that can interact with each group is deduced either by fragment-based interactions (hydrogen bonds, hydrophobic interactions, ionic interactions, etc.) or by interaction energy criteria. Once the fragments have been decided then a scaffold is chosen to connect them. The distance between the functional groups (Figure 3.15) is the criteria now to decide the scaffold.

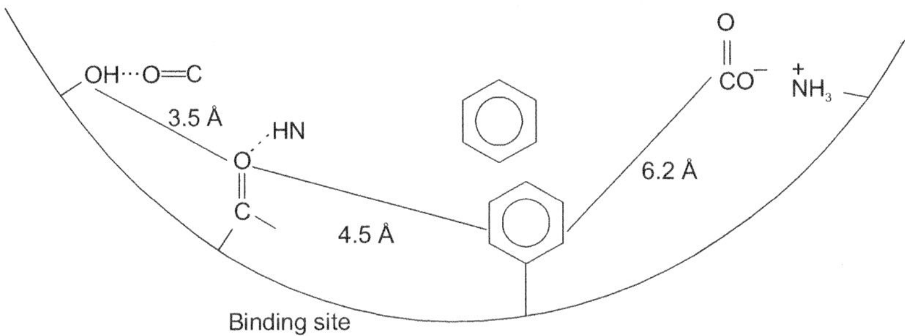

Figure 3.15 A generated pharmacophore

The structure with these pharmacophores can be searched in the database which might not yield the desired result. In such case, a chemist's intuition will help to synthesize the molecule. But the synthetic feasibility is one of the main reasons why de novo ligand design fails.

A possible structure-prediction would be as given in (Figure 3.16).

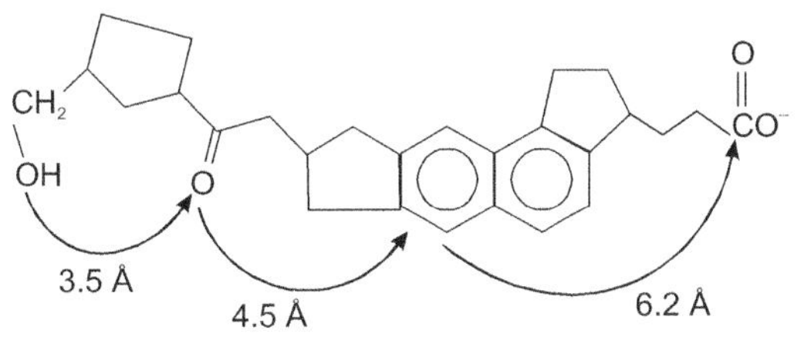

Figure 3.16 A virtual molecule based on the pharmacophore

But synthetically it would be an uphill task, as it would require multiple steps and the amount of percentage yield is questionable (if at all the drug is synthesized?). An ideal computational tool for this de novo lead design will be able to test *many* structures in a *short* period of time and arrange them into a ranked list based on an *accurate prediction of binding free energies,* since the latter reflect actual binding propensities.

REFERENCES

1. FDA (2004). Challenge and opportunities on the critical path to new medical products (http://www.fda.gov/oc/initiatives/criticalpath/whitepaper.html)

2. DiMasi, J.A. *et al.* (2003). "The price of innovation: New estimates of drug development costs." *J. Health Econ.* 22: 151–185.

3. McGee, P. (2005). "Modeling success with *in silico* tools." *Drug Discov. Today.* 8: 23–28.

4. Peter Imming, Christian Sinning and Achim Meyer. (2006). "Drugs, their targets and the nature and number of drug targets." *Nature Reviews, Drug Discovery.* Vol.5. pp. 821–834.

5. McCarter, J.P. (2004). "Genomic filtering: An approach to discovering novel antiparasitics." *Trends Parasitol.* 20: 462–468.

6. Geary, T.G., Thompson, D.P. and Klein, R.D. (1999). "Mechanism-based screening: Discovery of the next generation of anthelmintics depends upon more basic research." *Int.J.Parasitol.* 29: 105–112.

7. Schmidt, B.M., Ribnicky, D.M., Lipsky, P.E. and Raskin, I. (2007). *Nat. Chem. Biol.* 3: 360–366.

8. Timothy, W. Corson and Craig, M. Crews. (2007). "Molecular understanding and modern application of traditional medicines: triumphs and trials." *Cell.* 130: 769–774.

9. Jones,G., Willett, P., Glen,R.C., Leach, A.R.L. and Taylor, R. (1997). "Development and validation of a genetic algorithm for flexible docking." *Journal of Molecular Biology.* 267: 727–748.

10. Jones, G., Willett, P. and Glen, R.C. (1995). "Molecular recognition of receptor sites using a genetic algorithm with a description of desolvation." *Journal of Molecular Biology.* 245: 43–53.

11. Morris, G.M., Goodsell, D.S., Halliday, R.S., Huey, R., Hart, W. E., Belew, R. K. and Olson, A. J. (1998). "Automated docking using a lamarckian genetic algorithm and empirical binding free energy function." *J. Computational Chemistry.* 19: 1639–1662.

12. Jeffrey, S. Taylor and Roger, M. Burnett. "DARWIN: A program for docking flexible molecules." *Proteins: Structure, Function and Genetics.* Vol.41. 2: 173–191.

13. Zavodszky, M.I., Sanschagrin, P.C., Korde, R.S. and Kuhn, L.A. (2002). "Distilling the essential features of a protein surface for improving protein-ligand docking, scoring, and virtual screening." *J. Comp. Aided Molecular Design.* 16: 883–902.

14. McGann, M., Almond, H., Nicholls, A., Grant, J.A. and Brown, G. (2003). "Gaussian docking functions." *Biopolymers.* 68: 76–90.

15. Schellhammer and Rarey, M. (2004). "FlexX-Scan: Fast structure-based virtual screening." *Proteins: Structure, Function and Genetics.* 57: 504–517.

16. Welch, W., Ruppert, J. and Jain, A.N. (1996). "Hammerhead: fast, fully automated docking of flexible ligands to protein binding sites." *Chemistry & Biology.* Vol.3. 6: 449–462.

17. Eisen, M.B. and Wiley, D.C. *et al.* (1994). "HOOK: A program for finding novel molecular architectures that satisfy the chemical and steric requirements of a macromolecule binding site." *Proteins: Structure, Function and Genetics.* 19(3): 199–221.

18. Ewing, T.J.A., Makino, S., Skillman, A.G. and Kuntz, I.D. (2001). "DOCK 4.0: Search strategies for automated molecular

docking of flexible molecule databases." *Journal of Computer Aided Molecular Design.* Vol.15. 411–428.

19. Bursavich, M.G. and Rich, D.H. (2002). "Designing non-peptide peptidomimetics in the 21st century: Inhibitors targeting conformational ensembles." *J. Med. Chem.* 45: 541–58.

20. Leulliot, N. and Varani, G. (2001). "Current topics in RNA-protein recognition: control of specificity and biological function through induced fit and conformational capture." *Biochemistry.* 40: 7947–56.

21. Davis, A.M. and Teague, S.J. (1999). "Hydrogen bonding, hydrophobic interactions and failure of the rigid receptor hypothesis." *Angew Chem. Int. Ed. Engl.* 38: 736–749.

22. Venkatraman Mohan, Alan, C. Gibbs, Maxwell, D. Cummings, Edward, P. Jaeger and Renee, L. Desjarlais. (2005). "Docking: Successes and Challenges." *Current Pharmaceutical Design.* 11: 323–333.

23. Eldridge, M.D., Murray, C.W., Auton, T.R., Paolini, G.V. and Mee, R.P. (1997). "Empirical scoring functions: The development of a fast empirical scoring function to estimate the binding affinity of ligands in receptor complexes." *J. Comput. Aided Mol. Des.* 11: 425–45.

24. Rarey, M., Kramer, B., Lengauer, T. and Klebe, G. (1996). "A fast flexible docking method using an incremental construction algorithm." *J. Mol. Biol.* 261: 470–89.

25. Meng, E.C., Shoichet, B.K. and Kuntz, I.D. (1992). "Automated docking with grid-based energy evaluation." *J. Comput. Chem.* 13: 505–24.

26. Hart, T.N. and Read, R.J. (1992). "A multiple-start Monte Carlo docking method." *Proteins.* 13: 206–22.

27. Hart, T.N., Ness, S.R. and Read, R.J. (1997). "Critical evaluation of the research docking program for the CASP2 challenge." *Proteins. Suppl.* 1: 205–9.

28. Jones, G., Willett, P., Glen, R.C., Leach, A.R. and Taylor, R. (1997). "Development and validation of a genetic algorithm for flexible docking." *J. Mol. Biol.* 267: 727–48.

29. Muegge, I. and Martin, Y.C. (1999). "A general and fast scoring function for protein-ligand interactions: A simplified potential approach." *J. Med. Chem.* 42: 791–804.

30. Tame, J.R. (1999). "Scoring functions: A view from the bench." *J. Comput. Aided Mol. Des.* 13: 99–108.

31. Charifson, P.S., Corkery, J.J., Murcko, M.A. and Walters, W.P. (1999). "Consensus scoring: A method for obtaining improved hit rates from docking databases of three-dimensional structures into proteins." *J. Med. Chem.* 42: 5100–9.

32. Singh, A., Latombe, J. and Brutlag, D. "A motion planning approach to flexible ligand binding." (http://mgm.stanford.edu/~brutlag/Abstracts/singh98.html)

33. Goodsell, D., Morris, G. and Olson, A. (1996). "Automated docking of flexible ligands: Applications of AutoDock." *J. Mol. Rec.* 9: 1–5.

34. Nishio, M. *et al.* (1998). *The CH/π Interaction. Evidence, Nature and Consequences*. Wiley VCH, New York.

35. Desiraju, G. and Steiner, T. (1999). *The Weak Hydrogen Bond in Structural Chemistry and Biology*. Oxford University Press, Oxford.

36. Kubinyi, H. (2001). "Hydrogen bonding, the last mystery in drug design? pharmacokinetic optimization in drug research. biological, physicochemical, and computational strategies." Testa, B., van de Waterbeemd, H., Folkers, G. and Guy, R. (eds.). *Helvetica Chimica Acta*. Wiley-VCH, Zurich. pp. 513–524.

37. Abhishek Tiwari and Sunil, K. Panigrahi. (2007). "HBAT: A complete package for analysing strong and weak hydrogen bonds in macromolecular crystal structures." *In Silico Biology*. 7(6): 651–61.

38. Chen, J.M., Xu, S.L., Wawrzak, Z., Basarab, G.S. and Jordan, D.B. (1998). "Structure-based design of potent inhibitors of scytalone dehydratase: displacement of a water molecule from the active site." *Biochemistry.* 37(51): 17735–44.

39. Ladbury, J.E. (1996). "Just add water! The effect of water on the specificity of protein-ligand binding sites and its potential application to drug design." *Chem. Biol.* 3: 973–980.

40. Böhm, H.J. and Schneider, G. (eds.). (2000). *Virtual Screening for Bioactive Molecules.* Wiley-VCH.

41. Kramer, B., Rarey, M. and Lengauer, T. (1999). "Evaluation of the FLEXX incremental construction algorithm for protein–ligand docking." *Proteins.* 37: 228–241.

42. David, L., Luo, R. and Gilson, M.K. (2001). "Ligand–receptor docking with the mining minima optimizer." *J. Comput. Aided Mol. Des.* 15:157–171.

43. Diller, D.J. and Merz, K.M. Jr. (2001). "High throughput docking for library design and library prioritization." *Proteins.* 43: 113–124.

44. Ewing, T.J., Makino, S., Skillman, A.G. and Kuntz, I.D. (2001). "DOCK 4.0: Search strategies for automated molecular docking of flexible molecule databases." *J. Comput. Aided Mol. Des.* 15: 411–428.

45. Tondi, D., Slomczynska, U., Costi, M.P., Watterson, D.M., Ghelli, S. and Shoichet, B.K. (1999). "Structure-based discovery and in-parallel optimization of novel competitive inhibitors of thymidylate s ynthase." *Chem. Biol.* 6: 319–31.

46. Aronov, A.M., Munagala, N.R., Ortiz De Montellano, P.R., Kuntz, I.D. and Wang, C.C. (2000). "Rational design of selective submicromolar inhibitors of Tritrichomonas foetus hypoxanthine-guanine-xanthine phosphoribosyltransferase." *Biochemistry.* 39: 4684–91.

47. Perola, E., Xu, K., Kollmeyer, T.M., Kaufmann, S.H., Prendergast, F.G. and Pang, Y.P. (2000). "Successful virtual screening of a chemical database for farnesyl transferase inhibitor leads." *J. Med. Chem.* 43: 401–8.

48. Burkhard, P., Hommel, U., Sanner, M. and Walkinshaw, M.D. (1999). "The discovery of steroids and other novel FKBP inhibitors using a molecular docking program." *J. Mol. Biol.* 287: 853–8.

49. Wallace, A.C., Laskowski, R.A. and Thornton, J.M. (1995). "LIGPLOT: A program to generate schematic diagrams of protein-ligand interactions." *Prot. Eng.* 8: 127–134.

50. Orly Dym, Ioannis Xenarios, Hengming Ke and John Colicelli. (2002). "Molecular docking of competitive phosphodiesterase inhibitors." *Molecular Pharmacology.* Vol. 61. 1: 20–25.

51. Sheridan, R.P., Rusinko III, A., Ramaswany, N. and Venkataraghavan, R. (1989). "Searching for pharmacophores in large coordinate data bases and its use in drug design." *Proc.Natl. Acad. Sci.* USA. 86: 8165–8169.

Recommended Books for Reading

1. *Pharmacophore, Perception, Development and Use in Drug Design,* Edited by Osman, F.Guner, International University Line March 9, (2000). ISBN 0963681761.

2. *Virtual Screening: An Alternative or Complement to High Throughput Screening,* Edited by Gerhard Klebe, Kluwer Academic Publishers.(1999). ISBN 0-7923-6633-6.

3. *The Process of New Drug Discovery and Development*, Edited by Charles, G. Smith and James, T. O'Donnell, Informa Healthcare USA, Inc. (2006). ISBN 13: 978-0-8493-2779-7.

4. *Drugs From Discovery To Approval*, Rick Ng, John Wiley & Sons Inc. (2004). ISBN 0-471-60150-0.

5. *Drug Discovery Research New Frontiers in the post-genomic Era,* Edited By Ziwei Huang, John Wiley & Sons. (2007). ISBN 978-0-471-67200-5.

Websites of Related Interest

1. http://dock.compbio.ucsf.edu/

2. http://mgl.scripps.edu/

3. http://www.msg.ku.edu/~msg/MGM/links/dock.html

4. http://www.alzforum.org/drg/tut/tutorial.asp

5. http://www.chemlin.net/chemistry/drug_design.htm

Journals

1. *Chemical Biology & Drug Design* (Blackwell)

2. *Letters in Drug Design & Discovery* (Bentham)

3. *Journal of Computer-Aided Molecular Design* (Springer)

1. Natural products have been a source of lead molecules in the drug industry! Justify the statement.

2. Download the macromolecule 1pxx from Brookhaven data bank, remove the drug molecule from it, save it in the pdb format and using appropriate tool, convert it into mol2 format.

3. What is docking? Discuss the types of searching methods used in docking programmes, mentioning appropriate software for each method.

4. Is the binding of a drug to its receptor—an endothermic or exothermic process? Explain.

5. Can docking be termed as an *in silico* lead optimization process? Explain.

6. What is a scoring function? Briefly describe the types of scoring functions.

7. Comment on the statement, "The protonation states of the protein and the ligand, tautomeric forms and protein flexibility play a crucial role in the quality of the docking score."

8. What is the meaning of 'resolution' in X-ray crystallography? How does temperature affect the resolution?

9. Why is it very difficult to crystallize G-protein coupled receptors (GPCRs)?

10. There are different docking software like AUTODOCK, GOLD, Flexidock, etc. How is the efficacy of such docking tools determined? Does the AUTODOCK program consider the protonation states into effect.

11. In protein–ligand interactions, do you expect the agonist or the antagonist to possess less binding energy? Will their hydrogen bonding interaction pattern be the same or different?

4

CHEMOINFORMATICS

INTRODUCTION

Until recently, advances in medicinal and pharmaceutical chemistry depended on a trial-and-error-process aided by intuition. Though the properties that would indicate a certain molecule as a drug candidate were known, it was not really feasible to investigate large numbers of molecules for these types of properties. The nature of these properties would be represented by structural features of a molecule and thus examination of certain motifs provided a direction for experimental investigations. The problem with this approach is that it does not always lead to an understanding of why a molecule behaves as a drug against its target or why it does not. Furthermore, given a series of compounds, it is not always feasible to investigate experimentally which members of the series would be more potent or less toxic. As a result, though medicinal chemistry has resulted in a series of life-saving drugs, the process has traditionally been slow and tedious, and in many cases, advances have been due to serendipity rather than scientifically guided investigation.

In classical drug discovery, research had been based on vague hypotheses relating the structure with activity. Compounds were synthesized and tested in whole animals. If a biological effect was observed, a medicinal chemistry project started to optimize chemical structures with respect to activity, pharmacokinetic properties and lack of toxic side effects. This approach was modified to *in vitro* screening on defined targets, most often human proteins. Only in the past five years or so, more systematic investigation of druglike compounds in biological systems, termed as 'chemical biology' has been evolved.

The field of chemoinformatics is restricted to chemicals/drug molecules ignoring the protein information and forms an important tool in ligand-based drug design. Small molecules with atmost a few dozen atoms play a fundamental role in organic chemistry and biology. They can be used

◇ as combinatorial building blocks for chemical synthesis[1,2]

⬥ as molecular probes for perturbing and analysing biological systems in chemical genomics and systems biology[3,4]

⬥ for the screening, design and discovery of molecules with drug-likeliness[5,6]

Furthermore, huge arrays of new small molecules can be produced in a relatively short period of time using combinatorial approach.

The approaches to drug discovery in the field of chemoinformatics are based on the premise that knowledge of successful and failed ligand structures and properties are valuable for hit enrichment, appropriate lead identification and lead optimization. It is assumed that staying within the pre-defined/pre-discovered physico-chemical property ranges (like dipole moment, total surface area, etc.) where most successful compounds reside, and satisfying their structural features or similarities is very important. This will statistically enhance the chances of success in hit finding, identifying the best lead from multiple hits and generating the best candidate during lead optimization. Only 11% of drugs entering clinical development reach the market place, most are withdrawn from further studies mainly for reasons associated with efficacy, toxicity, drug metabolism and pharmacokinetics.

Chemoinformatics involves the application of information technology to chemical data and plays a key role in creation/ utilization of chemical databases, combinatorial library design, structure–activity relationships and structure-based drug design. Certain molecular features have been considered desirable in drug candidate molecules, such as molecular weights of <500, a limited number of H-bond acceptors <10, and donors < 5, and a clogP < 5, and these characteristics have been recognized in what have become known as the Lipinski "rule of five".

In chemoinformatics, the numbers represent the data, facts represent the information and rules represent the knowledge. Derivation of information and knowledge is only one aspect of chemoinformatics. The use of derived knowledge in a design and selection support role is an important part of the drug design cycle.

The 'real' world of compounds made in a chemistry laboratory and tested in a biological laboratory is only part of a much larger 'virtual' world where hypotheses may be computer-generated and tested for practicality. In this way, the expensive commitment to actual synthesis and bioassay is made only after exploring the initial concepts with computational models and screens.

WHAT IS CHEMOINFORMATICS?

Small molecules with atmost a few dozen atoms play a fundamental role in organic chemistry and biology. They can be used as combinatorial building blocks for chemical synthesis, as molecular probes for perturbing and analysing biological systems in chemical genomics and systems biology, and for the screening, design and discovery of useful compounds. These include drugs, the majority of which are small molecules. Furthermore, huge arrays of new small molecules can be produced in a relatively short period of time through combinatorial chemistry.

Chem(o)informatics is a generic term that encompasses the design, creation, organization, management, retrieval, analysis, dissemination, visualization and use of chemical information.[7] Chemoinformatics is the mixing of those information resources to transform data into information and information into knowledge for the intended purpose of making better decisions faster in the area of drug lead identification and optimization.[8] Chemoinformatics is the application of informatics methods to solve chemical problems.[9]

Chemoinformatics serves as an interface between chemistry and computer science, assisting in the drug design process. In this process, the representation of chemical structures is of critical importance. A hierarchy of descriptors for chemical structures has been developed. Descriptors are basically properties that describe the molecules. For example, dipole moment and hydrophobicity of a molecule fall under this category. The relationships between the structure of a molecule and its biological activity are too complex to be calculated directly by theoretical methods. In such situations, the modelling of the relationships between structure and properties by inductive learning methods such as statistical analyses, pattern recognition methods or neural networks offers an amenable solution.

The strength and the existence of chemoinformatics lies in the availability of open source academic research, open publications and open databases. The term "Open-source-software/database" connotes its availability in source-code form and licensed in a manner that permits unrestricted redistribution and creation of royalty-free derived works. Furthermore, the software license must not limit how, where or by whom the software can be used. Open source has already begun to emerge in drug discovery. Perl is widely used in bioinfomatics and Python is used in chemoinformatics and structural biology. Some notable examples include RasMOL [http:// www.openrasmol.org], Babel [http://www.openbabel.sf.net] and PyMOL [http://www.pymol.org]. However; most existing open-source scientific software have originated in academic settings.

Implementing, handling and searching chemical databases are a crucial aspect of chemoinformatics. Chemoinformatics tries to answer some of the fundamental questions of a chemist like searching a structure with a desired property, the synthetic feasibility of a compound, the possible product in a reaction, etc.

Using appropriate similarity measures, it might become possible to predict properties like Absorption, Distribution, Metabolism, Excretion and Toxicity (ADMET) at an earlier stage of the research pipeline, reducing expenditure per successful compound. In order to avoid animal testing, cosmetics and other consumer goods companies will focus on their in-house databases of chemical compounds that have already been tested for safety. Out of these compounds, some of them might already possess the desired properties, which could be detected by similarity searching.

Areas that come under chemoinformatics

◈ Structure/activity or structure/property relationships (QSAR, QSPR)

◈ Genetic algorithms

◈ Statistical tools (e.g. recursive pairing)

◈ Data analysis tools

◈ Visualization techniques

◈ Chemically-aware web language (CML)

CHEMICAL STRUCTURES AND REPRESENTATIONS

Since structure and sub-structure searching consume quite a bit of computer time, linear notations were introduced which convert structural graphs to strings that can easily be searched by a computer. The data screening strategies filtered out compounds that were not the main structural features (search keys) in a given query. The screens may include a range of possible hydrogen bond donors/acceptors, molecular weight range, number of aromatic rings, etc. Linear notations represent the structure of a chemical compound as a linear sequence of characters and numbers. The most popular linear notations are

Wiswesser Line Notation (WLN)[10]

In WLN, letters represent structural fragments and a complete structure is represented as a string. This system efficiently compressed structural data and was very useful in storing and searching chemical structures in low-performance computer systems. However, the WLN is difficult for non-experts to understand.

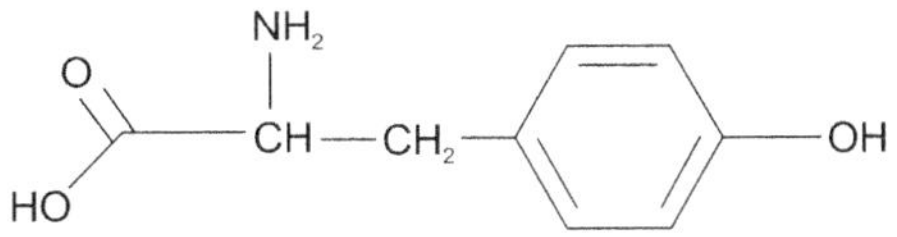

◈ WLN for the structure is **QVYZ1RDQ**

◈ Uses text symbolic representation of function groups.

e.g. $-Q$ = OH, **V** = –CO-, **Z** = –NH$_2$, **R** = benzene

◈ Other symbols represent branching, e.g. **Y**

Representation of Organic Structure Description Arranged Linearly (ROSDAL)

Structure diagram arranged linearly is represented as follows:

$$10-2=3O,2-4-5N,4-6-7=-12-7,10-3O$$

Simplified Molecular Input Line Entry Specification (SMILES)

SMILES notation[11] is very close to the 'natural language' used by organic chemists. SMILES is widely accepted and used in many chemical database systems. To successfully represent a structure, a linear notation should be canonicalized. That is, one structure should not correspond to more than one linear notation string and conversely, one linear notation string should only be interpreted as one structure.

◈ Possible SMILES for this structure is OC(= O)C(N)CC1 = CC =C(O)C = C1

The features of this notation include the following:

◈ There can be several ways of writing the same structure in SMILES, although a system of generating **canonical SMILES** exists.

❖ Atoms represented by their chemical symbol (C, N, S, O, Br, etc.)—uppercase for aliphatic, lowercase for aromatic.

❖ Adjacent atoms implicitly single bonded, or = for double bond, or # for triple bond.

❖ Hydrogens usually implicit. For example using SMILES notation, $CH_3-CH_2-CH_3-$ (propane) is represented as ccc.

❖ Parentheses represent branching.

❖ Ring enclosures represented by using numbers to signify attachment points

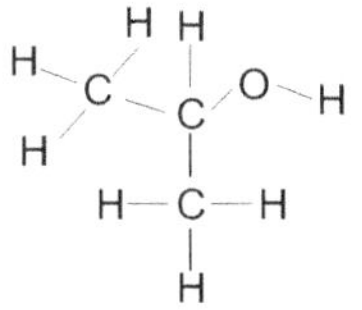

2-Propanol-CC(O)C

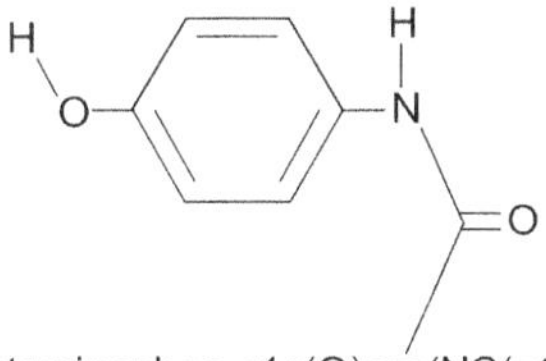

Acetaminophen-c1c(O)ccc(NC(=O)C)c1

SLN (SYBYL Line Notation)

$$OHC(=O)CH(NH_2)CH_2C[1]=CHCH=C(OH)CH=CH@1$$

Chemical databases and data sources

The multifaceted information about chemical compounds and reactions, such as literature data, physico-chemical properties, spectra, etc., can only be handled in a comprehensive manner by electronic methods. Chemical Abstract Service (CAS) registry (www.cas.org) is the largest and most comprehensive structure database (2D and predicted 3D structures), containing millions of records of organic and inorganic compounds, peptide sequences, proteins and nucleic acids. More than 30 million compounds are presently known in the Chemical Abstract Service entry. CAS is a division of the American Chemical Society and provides the world's largest and most current collection of chemical and related scientific information, including the most authoritative database of chemical substances, the CAS Registry. Its utilities include

1. Identifying patents to obtain information that has not been published otherwise, such as gene sequences, polymorphisms and experimental techniques

2. Finding compounds by searching the world's largest database of organic and inorganic compounds, including commercially available substances

3. Finding chemical reactions

4. Finding compounds based on compound properties like dipole moment, etc.

5. Finding enzymes based on CAS' extensive abstracting process, whereby enzymes with different names can still be found because they have been assigned the same CAS and EC numbers

CAS combines databases with advanced search and analysis technologies to yield a more complete, cross-linked and effective digital information environment for scientific research and discovery, including such products as SciFinder, SciFinder Scholar, STN, STN Express and STN AnaVist, among others. The amount of data and information constantly increases each year—more than one million new compounds and more than 70, 0000 publications that contribute in some way to chemical information. Hence, chemistry was one of the first scientific disciplines using databases to store its treasure of information. A competent overview on the topic can only be gained by querying databases and by data mining.

Types of Databases

Scientific databases can be normally categorized as

1. Literature (textual)

2. Factual (alphanumeric)

3. Structural (topological)

A strict separation into these categories and their subtypes is impossible, since many important chemistry databases cover several types of content. For instance, the Beilstein database contains structures, reactions, numerous physical properties of compounds and related literature references. Typical numerical databases are Beilstein, Gmelin, (covering inorganic and organometallic compounds), SpecInfo, KnowItAll, DETHERM, etc. Some of the most important catalog databases are the available Chemicals Directory from MDL, CHEMCATS and Chem Sources.

Representation and Searching of Chemical Structures

A diagrammatic representation of a compound is usually referred to as a structure diagram or structure graph. The structure diagram conventions are established as an international standard and so far are the *only unique* methods of communicating information on a chemical compound. The diagrams for computer processing are transformed (manually or automatically) into a linear string of characters or into two-dimensional matrices listing all the atoms (nodes) and their mutual interconnections (bonds, edges). These concise representations are referred to as connection tables (CT). The connection table consists of two parts: (1) some parameters common to a molecule, such as the name of the structure and (2) an array of atomic records which store a detailed description of each atom in the structure.

A chemical database can hence be represented by a large number of such graphs; with searching historically being carried out using two types of graphs or isomorphism algorithms. Structure searching involves an exact-match search of a chemical database for a specific query structure: this is required, for example, to retrieve the biological assay results and the synthetic details associated with a particular molecule. Such a search involves a graph isomorphism search, in which the graph describing the query molecule is checked for isomorphism

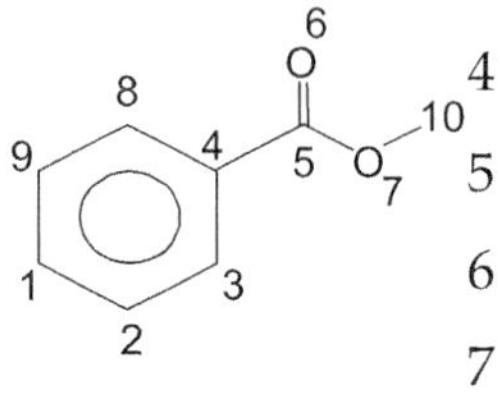

Index	Atomic number	Connected to atom	Bond type
1	6	2,9	4,4
2	6	1,3	4,4
3	6	2,4	4,4
4	6	3,5,8	4,1,4
5	6	4,6,7	1,2,1
6	8	5	2
7	8	5,10	1,1
8	6	4,9	4,4
9	6	1,8	4,4
10	6	7	1

(or structural equivalence) with the graphs of each of the database molecules. Substructure searching involves a partial-match search of a chemical database to substructure, irrespective of the environment in which that substructure occurs; for example, a user interested in antibiotics might wish to search a database to find all molecules that contain the characteristic penicillin ring nucleus. Table 4.1 gives an account of the available databases with their features.

Structure and Reaction Databases

Structure and reaction databases play a central role in chemical informatics, since they contain information on chemical structures, both as individual compounds and as participants in reactions. The structure diagrams are not stored as graphics (i.e., pictures which are not structure-searchable) but represented, e.g. as connection tables.

Table 4.1 Databases and their features

Database	Structural display	Sub-structure search	Number of compounds	Number added per year	Coverage	Experimental data
General database						
ACD	SD- Files, 2D, 3D	Yes	300,000			No
CAP[a] reagents	SD-Files, 2D, 3D[b]	Yes	239,219			No
SPRESI[web] 2.0	SD-Files, 2D	Yes	4,500,000		1974–2002[c]	Yes
Screening compounds						
MDL SCD	SD-Files, 2D, 3D	Yes	>2,000,000			No
SPECSnet[a]	SD-Files, 2D, 3D	Yes				
Database for medicinal agents						
CMC	SD-Files, 2D, 3D	Yes	8473	250	1900-present	Yes
MDDR	SD-Files, 2D, 3D	Yes	132,726	10,000	1988-present	No
WDI	SMILES, 2D, 3D	Yes	>73,000	3500	1983-present	No
NCI	2D, 3D		213,628			No
MedChem	SMILES 2D, 3D		48,500			Yes
WOMBAT	SMILES 2D, 3D		76,165		1992–2004	Yes

(Contd.)

Table 4.1 (Continued)

Database	Structural display	Sub-structure search	Number of compounds	Number added per year	Coverage	Experimental data
BIOSTER	SD-files, 2D, 3D	Yes	9500			No
CH/CRC Dictionary of Drugs		Yes	>41,000			Yes
Database with ADMET properties						
MDL Metabolite	SD-Files, 2D, 3D	Yes	>10,000[e]		1901–present	Yes
Accelrys' Metabolism[a]	SD-Files, 2D	Yes	4101[e]		1970–present	Yes
Accelrys' Biotransformation[a,d]	SD-Files, 2D	Yes	1744[e]		1987–1995	Yes
MDL Toxicity	SD-Files, 2D, 3D	Yes	>158,000		1902–present	Yes
DSS Tox	SD- Files, SMILES, 2D	No			2002– present	Yes
Databases with physico-chemical properties						
ACD/Labs physiochemical prop.[f]	SD-Files, 2D, 3D	No				Yes

Database	Structural display	Sub-structure search	Number of compounds	Number added per year	Coverage	Experimental data
PHYSPROP from SRC	SD-Files	Yes	25,250			Yes
AQUASOL dATAbASE		No	6000			Yes
CrossFire Beilstein	SD-Files, 2D, 3D		8,000,000			Yes
CRC Handbook			11,000			Yes

[a] Similarity search is also possible for these databases; [b] 3D display only possible in the version interfaced with the Catalyst program and not in the Accord version; [c] The Spresi database is being updated to present; [d] Accelrys' Biotransformation database is a subset in the Accelrys' Metabolism database; [e] Number of Parent compounds in the database; [f] A cluster of databases for a variety of properties, provided with a free graphical program, ACD/ChemSketch and can also be intefaced with ISIS/Base, ISIS/Draw and ChemDraw.

Structures include the topological arrangement of atoms and their connections as well as their stereochemistry. An example of structure database is the National Cancer Institute (NCI) database. Crystallographic structure databases, like the Inorganic Crystal Structure Database (ICSD), Cambridge Structural Database (CSD), and Protein Data Bank (PDB), provide "real" 3D structures from X-ray crystallographic structure analyses.

Patent databases including special structures are called Markush databases, e.g. MARPAT and Merged Markush Service. In reaction databases, structures of reaction participants are stored in a similar manner. Additionally, the role of each compound in the reaction (starting material, product, solvent, reagent, catalyst), the reaction centre information (formal, not mechanistic), and atoms added/ eliminated as well as bonds formed/broken in the reaction are stored. A large structure and reaction database is SPRESI (SPeicherung und REcherche Strukturchemischer Information; cp. ChemReact), Additional reaction databases, e.g. CASREACT, Chem-Inform, and Beilstein, contain only reaction information.

Search Methods in Chemical Informatics

Storing and searching chemical structures and associated information in databases forms the basis of chemoinformatics work. The majority of actual databases in chemistry contain chemical structures in computer-readable form, either 2D or 3D, connection tables or other representations. Therefore one of the primary methods for getting access to chemical information is searching for chemical structures or for a set of compounds that share a specific substructure or structural similarities. The structure representation is a major key in finding the appropriate results in a data set. Therefore the representation should include as much functionality of the structure as possible and it should also be unique and unambiguous. Various approaches have been devised for this purpose. They comprise the use of molecular formulae, molecular weights, trade and/or trivial names, various canonical line notations, registry numbers, constitutional diagrams (2D representations), atom coordinates (2D or 3D representations), topological indices, hash codes, etc. Rigorous methods of structure search (e.g. atom-by-atom) are computationally expensive. In particular, structure perception methods are used by consideration of items such as completeness, optimality, non-redundancy, and time and memory complexity.

Based on the problem or the need, a database can be searched simply for the atoms present or for stereochemical factors which requires a database of three-dimensional structures.

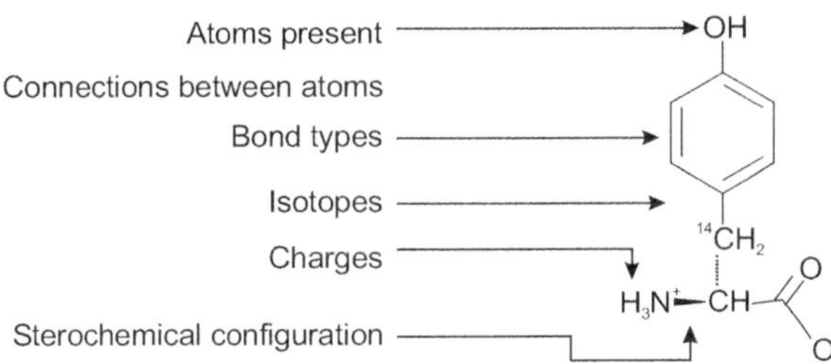

Substructure searching

Substructure searching is the process of identifying parts of a given structure that are equivalent to a specified query substructure. In graph theoretical terms, substructure searching verifies whether a query graph is isomorphic with the subgraph of another target graph by an atom-by-atom match algorithm, which represents a non-polynomial (NP) complete problem.[12] Representative algorithms are the Sussenguth algorithm (based on a partitioning procedure),[13] Figueras' algorithm (based on set reduction),[14] Ullmann's algorithm (based on backtracking[15] and relaxation refinements),[16] von Scholley's algorithm (based on relaxation),[17] and Xu's algorithms (based on backtracking and partial ordered sets).[18] Other important graph-based structure perceptions such as identification of equivalent atoms, determination of maximal common substructure, ring detection, the calculation of topological indices, etc. may also be comprised in substructure search algorithms.[19]

Searching a database for a given structure (Query) using the substructure options would yield compounds with this unit as its component. Given below is the illustration of the results of searching a database for diphenyl ether. Two dimensional searches do not yield any information regarding stereochemistry, chirality, etc., since they are the attributes of the third dimension.

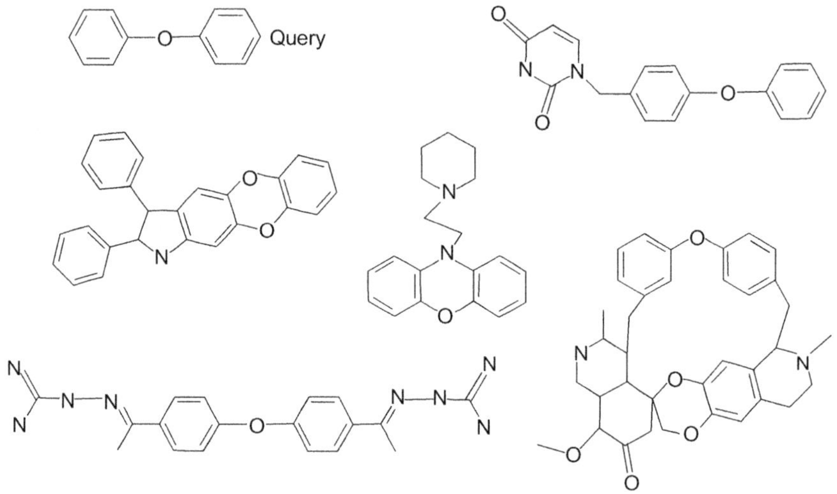

CASE STUDY

Using PUBCHEM for substructure search of the compound-diphenyl ether

The pubchem (http://pubchem.ncbi.nlm.nih.gov/search/search.cgi) can be searched based on any one of the following

STEP 1

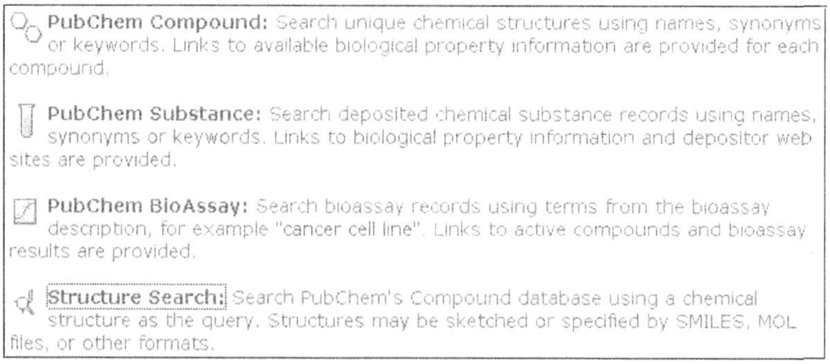

PubChem Compound: Search unique chemical structures using names, synonyms or keywords. Links to available biological property information are provided for each compound.

PubChem Substance: Search deposited chemical substance records using names, synonyms or keywords. Links to biological property information and depositor web sites are provided.

PubChem BioAssay: Search bioassay records using terms from the bioassay description, for example "cancer cell line". Links to active compounds and bioassay results are provided.

Structure Search: Search PubChem's Compound database using a chemical structure as the query. Structures may be sketched or specified by SMILES, MOL files, or other formats.

STEP 2

The compound can be drawn and can be searched through substructure option.

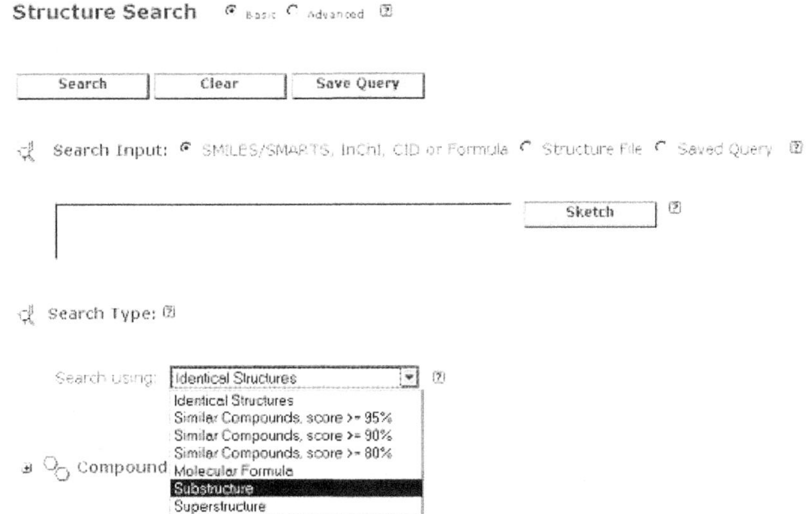

STEP 3

The diagram can be drawn in the panel that would appear, as shown below.

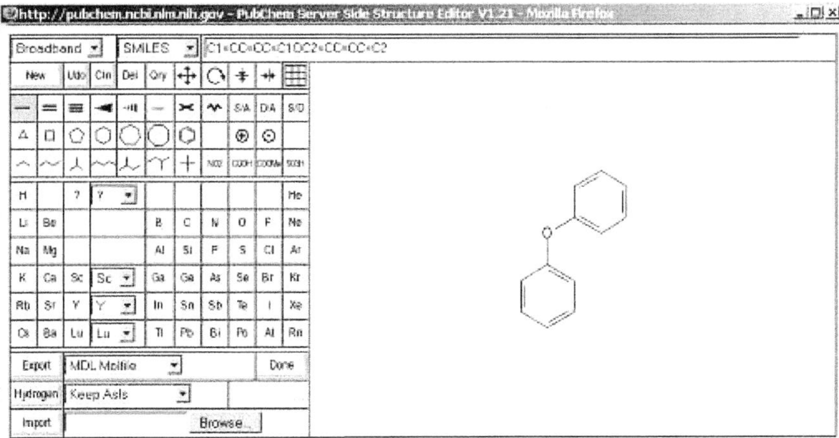

STEP 4

The structure is converted to the SMILES format on clicking the "Done" button.

Result

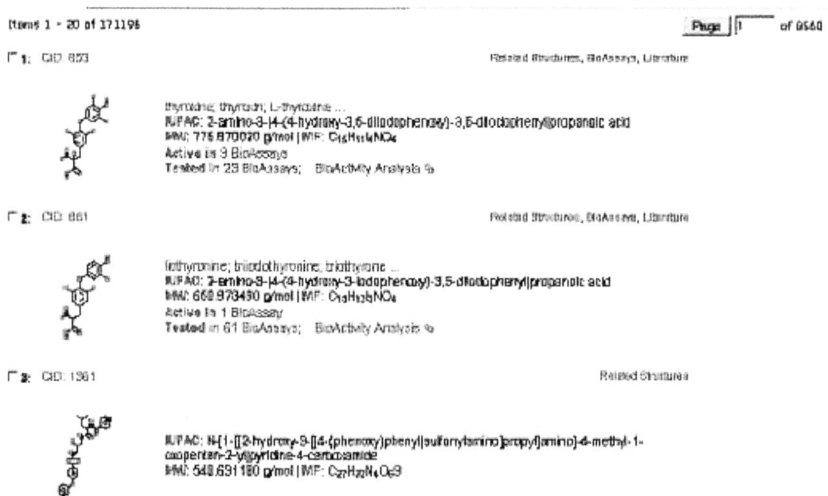

How to use these results (hits) in drug design   The compounds that have been obtained can be as such used for virtual screening (by docking), which requires advanced/commercial software.

Otherwise the compounds can be filtered based on the properties (which can be the clogP range, number of rotatable bonds, etc.) of the known compounds. The remaining compounds can then be docked and categorized based on their scoring function.

Similarity searching

Substructure search is appropriate for highly constrained database querying. The specification of a certain substructure is quite often insufficient for the definition of a family of related structures. Occasionally one wants to obtain a group of structures that have a number of structural features in common: structures that are similar to each other. A similarity search compares a set of characteristics describing the target structure with the corresponding sets of characteristics for each of the database structures. The measure of similarity between the target and each database structure is calculated based on the degree of resemblance of these two sets of characteristics. The database structures are usually sorted into order of decreasing similarity with the target.

The most important similarity measure has three main components: the structural representation that is used to characterize the molecules; the weighting scheme that is used to differentiate more important features from less important features; and the similarity coefficient that is used to quantify the degree of similarity between pairs of molecules. These coefficients may be based on fragments (e.g. the Tanimoto coefficient), on topological indices (and other calculated physico-chemical properties, e.g. molecular connectivity, dipole moment, etc.) and on graphs (e.g. maximum common subgraphs).

The coefficients for computing the similarity between two fingerprints are the following.

$$\text{Tanimoto coefficient } \frac{c}{a+b-c}$$

$$\text{Cosine coefficient } \frac{c}{\sqrt{ab}}$$

$$\text{Hamming distance } a + b - 2c$$

Each coefficient computes the similarity between two molecular fingerprints, X and Y, of length n, in which a is the number of bits set in both X and Y, b is the number of bits set exclusively in X, c is the number of bits set exclusively in Y (so that $n = a + b + c$)

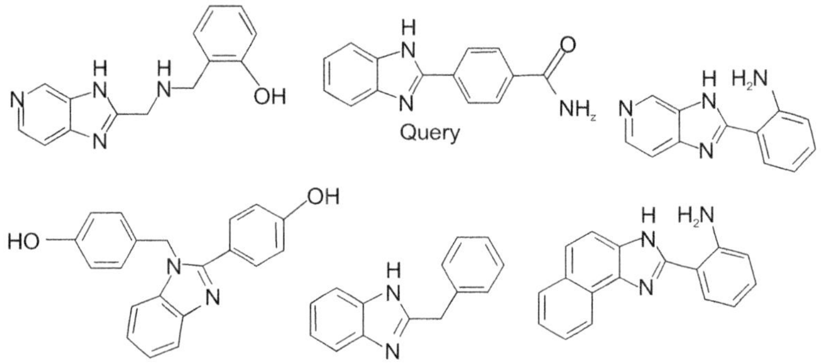

Query

Given below is the search protocol for the above query in PUBCHEM

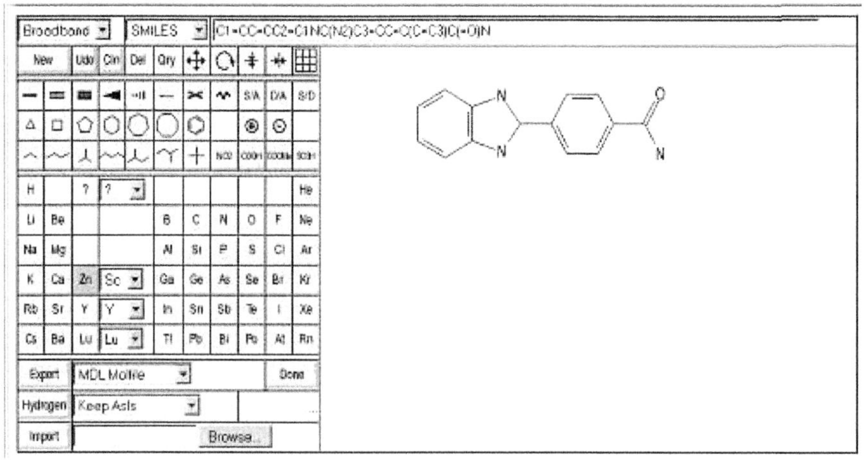

C1=CC=CC2=C1NC (N2) C3=CC=C (C=C3) C (=O) N

♂ Search Type: ⑦

Search using: [Similar Compounds, score >= 95% ▾] ⑦

The search with Score >= 95% yielded no results

Score >= 90% yielded 41 structures

Score >= 80% yielded 7841 structures

Searching the NCI database—an example

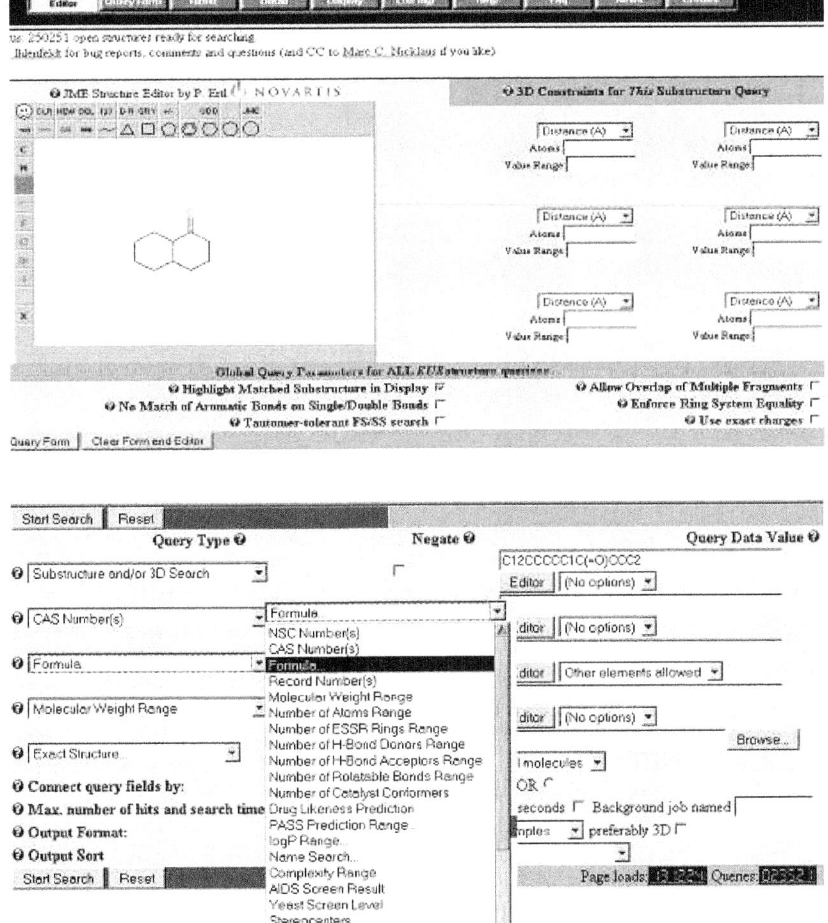

	Sample Structures			
2976	5425	5682	8640	9132

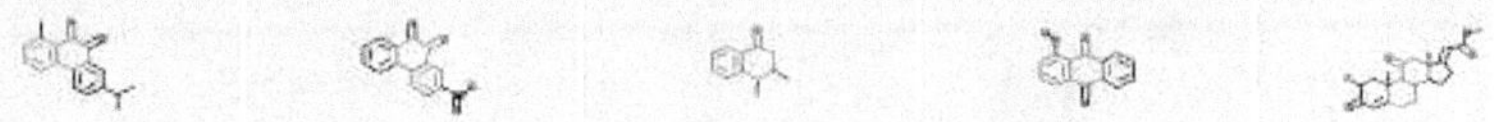

	NSC Number	Formula	CAS	#Names	Sample Name
☑	5	$C_{14}H_9NO_2$	117-79-3	10	2-aminoanthra-9,10-quinone
☑	8	$C_{15}H_9NO_4$	129-15-7	8	1-(hydroxy(oxido)amino)-2-methylanthra-9,10-quinone
☑	201	$C_{14}H_8O_8S_2$	117-14-6	5	9,10-dioxo-9,10-dihydro-1,5-anthracenedisulfonic acid
☑	202	$C_{14}H_8O_8S_2$	82-48-4	4	9,10-dioxo-9,10-dihydro-1,8-anthracenedisulfonic acid
☑	458	$C_{14}H_9NO_2$	82-45-1	10	1-aminoanthra-9,10-quinone
☑	504	$C_{15}H_9ClO_2$	129-35-1	4	1-chloro-2-methylanthra-9,10-quinone
☑	607	$C_{15}H_{10}O_2$	84-54-8	8	2-methylanthra-9,10-quinone
☑	1485	$C_{14}H_9NO_3$	116-85-8	44	1-amino-4-hydroxyanthra-9,10-quinone
☑	1613	$C_{27}H_{44}O_4$	5346-42-9	1	3-hydroxy-4,4,14-trimethyl-11-oxocholan-24-oic acid

NSC Number:	201	Date:	2008-04-14 03:56
File Record:	201	CAS Number:	117-14-6
Formula:	$C_{14}H_8O_8S_2$	Weight:	368.3324 g/mol
Complexity:	634.3	Anti-HIV Screening:	*No data available*
Druglikeness(std):	*No drug*	logP(KOW):	-0.76
Druglikeness(nog):	*Is drug*	logP(exp):	*No data*
WDI Record:	*No*	logP(ACD):	*No data*
H-Bond Acceptors:	8	Available on DTP Plates:	Yes
H-Bond Donors:	2		
# Rotatable Bonds: (CACTVS)	2	WLN:	*No data*
Stereochemistry Potential *R/S atoms and E/Z bonds*	N_0	Yeast Screen Level	0
# Catalyst Conformers: *(0 if Catalyst could not handle structure)*	6	Matched Conformer:	*None*

C 45.65% H 2.19% O 34.75% S 17.41%

OS(=O)(=O)C1=C2C(=O)C3=CC=CC(=C3C(=O)C2=CC=C1)S(O)(=O)=O

9,10-dioxo-9,10-dihydro-1,5-anthracenedisulfonic acid (ACD/Name 4.0)
Anthraquinone-1,5-disulfonic acid

	1,5-Disulfoanthraquinone
Commercial Availability:	Asinex, ACD
Commercial Database Keys:	acd:MFCD00019154, asinex:BAS_0073376
Available Screening Data:	*No screen data available*
Anti-HIV Screening:	*No data (EC50/IC50) available.*
Cancer Screening Summary:	*No data (GI50/TGI/LC50) available.*

PASS Predictions: Predicted Activity	p(active)	p(inact)
Acetylcholine release stimulant	0.367	0.059
ADP ribosyl transferase inhibitor	0.287	0.071
Adrenergic transmitter uptake inhibitor	0.410	0.107
Alpha 1 adrenoreceptor agonist	0.249	0.152
Alpha 2 adrenoreceptor agonist	0.176	0.028
Alpha adrenoreceptor agonist	0.111	0.110
Alzheimer's disease treatment	0.326	0.266
Aminopeptidase microsomal inhibitor	0.326	0.091

3D substructure searching

Structure diagrams are planar but molecules are not, three-dimensional structures can be obtained either from experimental data (Cambridge Structure Database), Computational chemistry (Quantum Mechanics, Molecular Mechanics, and Molecular Dynamics) or by structure-generation/2D to 3D conversion using software such as CONCORD and CORINA.

DESCRIPTORS, PHYSICAL AND CHEMICAL DATA

One major topic in chemoinformatics is how chemical structure information can be correlated with physical, chemical or biological data and how such a model, once established, can be used for the prediction of new data. Thus, the appropriate description of chemical structures for the property to be modelled is an essential task. With a multitude of different types of descriptors being available (more than 1500), it is important in identifying suitable set of descriptors. Quantitative structure–activity relationships (QSAR) or Quantitative structure–property relationships (QSPR) are usually established by inductive learning methods.

In inductive learning methods, one learns from observations, from data. These data are put into context to obtain information. Information can then be generalized to obtain knowledge.

To give an example, the measurement of a certain biological activity is, by itself, not very useful. Only when we can associate such a biological activity with a chemical structure do we obtain information. Many such pieces of information of chemical structures and their associated biological activities can then be used to build a model for the relationships between chemical structure and biological activity. Such a model comprises knowledge that can be used to make predictions on the biological activity of new chemical structures.

In recent decades, methods have been developed that allow inductive learning to be put on a more formal and rigorous basis by mathematical methods. This includes areas of machine learning, data mining, pattern recognition, chemometrics, neural networks, etc. All these methods are considered to be part of chemoinformatics.

In deductive learning, a theory is used to make inferences and deductions. In Chemistry, this is usually achieved by calculations such as quantum mechanical or molecular mechanical calculations. Such calculations provide data that can assist in solving a problem.

An essential requirement for such learning methods to be invoked is that all structures of a data set are represented by the same, fixed number of descriptors. Thus, the representations of 2D structures (connection tables) or 3D structures (e.g. Cartesian coordinates) cannot be used as such since the number of descriptors would depend directly on the number of atoms in a molecule; more the atoms, more are the descriptors. Therefore, chemical structure information has somehow to be mathematically transformed to provide for any molecule descriptors, irrespective of the size of a molecule and of the number of atoms in a molecule. Various structure coding methods mainly developed in Gasteiger's research group take account of various physico-chemical effects and can be scaled either to consider only the constitution of a molecule/the three-dimensional structure/ molecular surface properties.

There are reasons that make chemoinformatics indispensable: the amount of information available in chemistry is enormous. Presently, more than 40 million different compounds are known; all have a series of properties (physical, chemical or biological), can be synthesized in many different ways, made by a wide range of reactions and characterized by a host of spectra. Each year more than a million new compounds are discovered or synthesized. Each year about 800,000 new articles are published that somehow deal with aspects of chemistry. All this just aggravates the flood of information. This immense amount of information can only be processed by electronic means, by the power of the computer.

Structure Descriptors and Profiling Compound Libraries

After the pharmaceutical industry adopted high throughput techniques in the 1990s, quick profiling of a compound library with thousands or millions of chemical structures became an important issue. The purpose of profiling compound libraries is to answer the following questions:

1. How diverse is a library?

2. How similar are compounds in the corporate library compared with marketed drugs?

3. How should one select a sub-library that structurally represents the whole library?

4. Is a foreign library structurally complementary to the corporate library?

Consideration of these questions is known as diversity analysis. In order to do diversity analysis, in Chemoinformatics, a high throughput data mining approach is employed. Currently, structure descriptors are used not only as sub-structure bit-maps, but also to represent any structural property. These can be based upon topological or three-dimensional properties such as molecular indices, molecular weight, number of H-bond donors, etc. The tools to calculate various structure descriptors are available publicly and commercially. Structure descriptors are fundamental tools to profile compound libraries, and diversity analysis is one of the main components of modern chemoinformatics.

Since there are many structure descriptors, it is important to identify the redundant ones and those that are correlated with each other, as too many descriptors will increase computation costs. If a structure is represented by two or three structure descriptors, a compound library containing thousands of structures can be graphed in a two- or three-dimensional space using the descriptors as coordinates. Such graphs visually show compound structural diversity of a database. However, it should not be one's objective to represent a structure just using three descriptors. It would be better to represent a structure with many (perhaps one hundred) descriptors. However, diversity graphs would then have to be one hundred-dimensional. In order to view a one hundred-dimensional space, one needs a technology to project higher dimensional data space to two- or three-dimensional space. This technology is known as dimension reduction.

Principal component analysis (PCA) and factor analysis (FA) are usually used to filter out redundant descriptors and eliminate descriptors having minor information contribution. PCA is used to transform a number of potentially correlated variables (descriptors) into a number of relatively independent variables that then can be ranked based upon their contributions for explaining the whole data set.

Principal Component Analysis

Principal Component Analysis is a variable reduction procedure. It is useful when you have obtained data on a number of variables (possibly a large number of variables) and believe that there is some redundancy in those variables. In this case, redundancy means that some of the variables are correlated with one another, possibly because they are measuring the same construct. Because of this redundancy, you believe that it should be possible to reduce the observed variables into a smaller number of principal components (artificial variables) that will account for most of the variance in the observed variables.

PCA is used abundantly in all forms of analysis—from neuroscience to computer graphics—because it is a simple, non-parametric method of extracting relevant information from confusing data sets. With minimal additional effort, PCA provides a roadmap for how to reduce a complex data set to a lower dimension to reveal the sometimes hidden, simplified structure that often underlie it.

APPLICATIONS OF CHEMOINFORMATICS IN DRUG DISCOVERY

Compound Selection

For many pharmaceutical organizations, HTS capacity is allocated on two levels: the number of targets screened and the number of samples screened per target. Screening all available compounds against all available targets is beyond the HTS capacity of pharmaceutical organizations. On the other hand, for a given parallel synthesis protocol and for available reactants, combinatorial chemistry can make huge numbers of compounds which are, as well, beyond the HTS capacity of pharmaceutical organizations. Therefore, one must apply some method to select a smaller set of compounds from a large compound pool. The main tasks for compound selection are

1. To select and acquire compounds from external sources that will provide complementary diversity to existing libraries

2. To select for screening, from a corporate compound pool, a subset that provides diversity representation

3. To select reagents to make a combinatorial library which will maximize diversity and

4. To select compounds, from available compound collections that are similar to known ligands, with different and novel scaffolds. Diversity-based compound selection has been done using many classification approaches.

In Silico ADMET

For any drug development effort, oral bioavailability is often a requirement to compete in the marketplace with drugs already available. It has been estimated that roughly 10% of the compounds that enter development eventually become marketed drugs and 40% of compounds fail due to poor pharmacokinetic properties.[20]

The ability to predict so called ADMET (Absorption, Distribution, Metabolism, Excretion and Toxicology) properties from molecular structure has a tremendous impact on the drug discovery process both in terms of cost and the amount of time required to bring a new compound to market.

Methods for estimating the molecular properties that correlate with ADMET problems have also been a very active arena for Chemoinformatics. By studying the isozymes of cytochrome P450 enzymes, for example, certain molecular signatures for metabolic stability have been discerned.[21] In a similar way, properties such as lipophilicity, pKa, number of hydrogen-bond donors and acceptors, and so on, correlate with oral bioavailability.

For a bioactive compound to succeed as a drug, it must pass many selective filters during development (toxicity, etc.) as well as in the body including metabolism, uptake, and excretion and so on.

Historically, drug absorption, distribution, metabolism, excretion and toxicity (ADMET) studies in animal models were performed after a lead compound was identified. Now, pharmaceutical companies are employing higher throughput, *in vitro* assays to evaluate the ADMET characteristics of potential leads at earlier stages of development. This is done in order to eliminate candidates as early as possible, thus avoiding costs, which would have been expended on chemical synthesis and biological testing. Scientists are developing computational methods to select only compounds with reasonable ADMET properties for screening. Molecules from these computationally screened virtual libraries can then be synthesized for high-throughput biological activity screening. As the predictive ability of ADME/Tox software improves and as pharmaceutical companies incorporate computational prediction methods into their R&D programs, the drug discovery process will move from a screening-based to a knowledge-based paradigm. Under multi-parametric optimization drug discovery strategies, there is no excuse for failing to know the relative solubility and permeability rankings of collections of chemical compounds for lead identification.

Absorption

Passive Intestinal Absorption (PIA) models have been studied by many groups, for years. The fluid mosaic model holds that the structure of a cell membrane is an interrupted phospholipid bilayer capable of both hydrophilic and hydrophobic interactions.[21] Trans-cellular passage through the membrane lipid/aqueous environment is the predominant pathway for passive absorption of lipophilic compounds, while low-molecular-weight (< 200), hydrophilic compounds make use of the water-filled channels of the tight junctions between membrane cells (paracellular transport).[22] Therefore, lipophilicity is considered a key property for activity in drug design and is a common property used to estimate the membrane permeability of a molecule. Lipophilicity is measured as the log of the partition coefficient between *n*-octanol and water (logP). logP prediction programs are available and results are reasonably good.[22–24]

But, the relationship between logP and permeability is not linear. Permeability drops at both low and high logP. It is theorized that these non-linearities are due to

1. The inability of weakly lipophilic compounds to penetrate the lipid portion of the membrane

2. The excessive partitioning of strongly lipophilic compounds into the lipid portion of the membrane and their subsequent inability to pass through the aqueous portion of the membrane. A strong relationship between passive intestinal absorption and polar surface area (PSA) has been discovered by several groups.[25–28]

Also linear and non-linear multivariate models have been introduced to model PIA based upon logP, molecular weight, H-bonding, free energy, H-bond donor, H-bond acceptor, polarizability, numbers and strengths of H-bond acceptor nitrogen and oxygen atoms, number of H-bond donor atoms and lipophilicity (log D at pH 7.4) on the Caco-2 cell permeability. Genetic algorithm is commonly used to select the best descriptors for predictive models.

Distribution

Drugs acting on central nervous system must cross the blood-brain barrier (BBB). The experimental determination of the brain-blood partition ratio is difficult and time-consuming to compute since it involves the direct measurement of the drug concentration in the brain and blood of laboratory animals. This obviously requires the synthesis of the compounds, often in radio-labelled form. *In vitro* techniques to predict brain penetration are available, but they are experimentally cumbersome. The earlier work involved in correlating log (C_{brain}/C_{blood}) or logBB and logP (octanol-cyclohexane), $P_{cyclohexane}$, or $logP_{oct}$, which basically reflect the polar nature of molecules was based upon smaller (about 20 compounds) data sets.[29, 30]

More descriptors have been correlated with logBB, such as excess molar refraction, solute polarizability, hydrogen bond acidity and basicity, and molecular volume. Descriptors derived from 3D molecular fields to estimate the BBB permeation on a larger set of compounds and to produce a simple mathematical model have been studied. The method used (VolSurf) transforms 3D fields into descriptors and correlates them to the experimental permeation by a partial least squares procedure.[31]

Human serum albumin (HSA) protein is the major transporter of non-esterified fatty acids, as well as of different drugs and metabolites to different tissues. HSA allows solubilization of hydrophobic compounds, contributes to a more homogeneous distribution of drugs in the body and increases their biological lifetime. The binding strength of any drug to serum albumin is the main factor for availability of that drug to diffuse from the circulatory system to target tissues. All these factors cause the pharmacokinetics of almost any drug to be influenced and controlled by its binding to serum albumin.[32]

Therefore, QSAR study on binding of drugs and metabolites to HSA is extremely important for drug distribution. Biosensor analysis for prediction of HSA has been reported. In order to build an *in silico* predictive model for binding affinities to HSA, Colmenarejo and co-workers at GlaxoSmithKline used a genetic algorithm to exhaustively search and select for multivariate and non-linear equations, starting from a large pool of molecular descriptors. They found that hydrophobicity (as measured by the clogP) is the most important variable for determining the binding extent to HSA. Binding to HSA turns out to be determined by a combination of hydrophobic forces together with some modulating shape factors.[33] This agrees with X-ray structures of HSA alone or, bound to ligands, where the binding pockets of both sites I and II are composed mainly of hydrophobic residues.[34]

Metabolism

Drug metabolism is another barrier to overcome. Metabolism is studied, by *in vitro, in vivo* and *in silico* approaches. *In vitro* approaches determine metabolic stability, screening for inhibitors of specific cytochrome P450 isozymes and, identifying the most important metabolites. *In vivo* approaches measure hepatic metabolic clearance, volume of distribution and bioavailability, and identify major metabolites. *In silico* approaches are categorized into three classes[35] — QSAR and pharmacophore models, protein models, and expert systems.

QSAR and pharmacophore models predict substrates and inhibitors of a specific cytochrome P450 isozyme. Protein models rationalize metabolite formations and identify possible substrates, potential metabolites or inhibitors by means of docking algorithms. Therefore, they introduced a metabolic fingerprint concept, METAPRINT, for the assessment of metabolic similarity and diversity in combinatorial chemical libraries. Their metabolic fingerprint was developed by predicting metabolic pathways and corresponding potential metabolites.

Excretion/elimination

Certain drugs like Non-Steroidal Anti-Inflammatory compounds (NSAIDs) such as Ibrufen stay in the body for longer time causing unnecessary complications. Recently, Nimesulide (a NSAID) has been banned in certain countries for inducing liver injury and fatal hepatotoxicity and neonatal renal failure.[36, 37]

In such cases, the prediction of their half-life, which determines the length of time a drug will persist in the body, is important in order to reduce subsequent drug failures. Prediction of half-life is difficult, due to the multi-faceted nature of drug elimination. Distribution of drug in fat and major organs, excretion by kidneys and metabolism by liver all contribute to the rate at which a drug is eliminated from the body. On the other hand, it may be possible to make use of qualitative predictions of half-life. Such information can be used, for example, to predict whether a drug is likely to accumulate to a significant extent when used for prolonged treatment.[38]

Toxicity

Current toxicity prediction approaches use either mechanistic or correlative methods. Correlative systems take molecular descriptors, biological data and chemical structures and by use of statistical analysis of data sets, represent them in mathematical models. The models describe the relationships between structure and activity and can be used to predict toxicity. The mechanistic approach involves human experts who make a considered assessment of the mechanism of interaction with a biological system, taking the molecular properties, biological data and chemical structures into account.[39]

The correlative approach uses an unbiased assessment of the data to generate relationships and predict toxicity. It is capable of discovering potentially new SARs [40]and, can lead to new ideas in the human assessment of mechanisms by which chemicals interact with biological systems.

A list of organizations providing ADMET solutions is given in Table 4.2.

Table 4.2   List of software for predicting the physiochemical properties

Software	Availability	Website address
Absolv-2	Purchase	www.ap_algorithms.com
ACD/Labs[a]	Purchase	www.acdlabs.com
Admensa	Purchase	www.inpharmatica.com
ADME Boxes	Purchase	www.ap_algorithms.com
ADMET Predictor[a]	Purchase	www.simulationsplus.com
ASTER[b]	Not available for public use	www.ena.gov
ChemAxon	Purchase	www.chemaxon.com
ChemOffice	Purchase	www.cambridgesoft.com
ChemProp	Not Known	www.ufz.defindex.php?en=6738
ChemSilico	Purchase[c]	www.chemsilico.com
ClogP	Purchase	www.daylight.com
Episuite	Freely downladable	www.epa.gov/oppt/exposure/pubs/episuitedl.htm

(Contd.)

Table 4.2 (Continued)

Software	Availability	Website address
Molecular Modeling Pro	Purchase	www.chemsw.com
Pallas	Purchase	www.seitegic.com
Pipeline Pilot	Purchase	mwsoftware.com/dragon/
ProPred	Consortium members only	www.capec.kt.dtu.dk
QikProp	Purchase	www.schrodinger.com
SPARC[d]	Free on-line	Ibmlc2.chem.uga.edu/sparc
TSAR	Purchase	www.accelrys.com
VCCLAB	Free on-line	www.vcclab.org

[a] Aqueous solubility module predicts intrinsic solubility solubility in pure water and solubility at user-specified pH; [b] Aster is currently not available for public use. It is hoped that it will at some point be available on www. It is understood that OECD may also add it to their QSAR Toolbox; [c] Log K_{ow} and aqueous solubility predictions available free on line at www.logp.com, but not in batch mode; [d] Not available in batch mode

Failures to Accurately Predict the Properties

In spite of all the computational advances, the predictive models do have quite a bit of noise which questions the dependability of the prediction. From the point of view of molecular modelling and computational chemistry, the potential functions in common use have intrinsic significant errors in electrostatics. Estimating the entropy of binding is complex unless one is willing to sample solvent configurations sufficiently to adequately represent the partition function. Solvation models such as Generalized Born/Surface Area GB/SA, which take solvent effect into consideration (which needs to be), are certainly better than ignoring the significant impact that desolvation has on energetics. Multiple binding modes are not uncommon but often too difficult to handle while modelling. The normal assumption of rigid receptor sites, or at the very least, limited exploration of the dynamics of the structure seen in the crystal, is inherently misleading. Receptors, at least the G-Protein Coupled Receptors, have multiple conformations and probably different modes of activation and coupling with different G-proteins. The role of dimerization of GPCRs has only recently been shown to be important for a variety of receptors. Such factors do affect the quality of the research if ignored.

REFERENCES

1. Schreiber, S.L. (2000). "Target-oriented and diversity-oriented organic synthesis in drug discovery." *Science.* 287: 1964–1969.

2. Agrafiotis, D.K., Lobanov,V.S. and Salemme,F.R. (2002). "Combinatorial informatics in the post-genomics era." *Nature Reviews Drug Discovery.* 1: 337–346.

3. Stockwell, B.R. (2004). "Exploring biology with small organic molecules." *Nature.* 432: 846–854.

4. Dobson, C.M. (2004). "Chemical space and biology." *Nature.* 432: 824–828.

5. Lipinski, C. and Hopkins, A. (2004). "Navigating chemical space for biology and medicine." *Nature.* 432: 855–861.

6. Jonsdottir, S.O., Jorgensen,F.S. and Brunak,S. (2005). "Prediction methods and databases within chemoinformatics: Emphasis on drugs and drug candidates." *Bioinformatics.* 21: 2145–2160.

7. Houghten, R.A. (2000). "Parallel array and mixture-based synthetic combinatorial chemistry: Tools for the next millenium." *Annual Review of Pharmacology and Toxicology.* 40: 273–282.

8. Paris, G. (August 1999 ACS meeting), quoted by Warr,W.A. at http://www.warr.com/warrzone.htm.

9. Brown, F.K. (1998). "Chemoinformatics: What is it and how does it impact drug discovery?" *Ann. Reports Med. Chem.* Vol. 33. pp. 375–384.

10. Gasteiger, J. and Engels, T. (2003). *Chemoinformatics: A Textbook.* Wiley-VCH.

11. William, J. Wiswesser. (1982). "How the WLN began in 1949 and how it might be in 1999." *J. Chem. Inf. Comput. Sci.* (2): 88–93.

12. Weininger, D. (1988). "SMILES, a chemical language and information system." *J. Chem.Inf.Comput Sci.* 28: 31–36.

13. Garey, M. and Johnson, D. (1979). *Computers and Intractability: A Guide to the Theory of NP-Completeness.* W.H. Freeman & Co. p. 340.

14. Barnard, J.M. (1993). "Substructure searching methods: Old and new." *J. Chem. Inf. Comput. Sci.* 33: 532–538.

15. Sussenguth, E.H. (1965). "A graph-theoretic algorithm for matching chemical structures." *J. Chem. Doc.* 5: 36–43.

16. Figueras, J. (1972). "Substructure search by set reduction." *J. Chem. Doc.* 12: 237–244.

17. Ray, L.C. and Kirsch, R.A. (1957). "Finding chemical records by digital computers." *Science.* 126: 814–819.

18. Bayada, D.M., Simpson, R.W, Johnson, A.P. and Laurenco, C. (1992). "An algorithm for the multiple common sub graph problem." *J. Chem. Inf.Comput. Sci.* 32 (6): 680-5.

19. Von Scholley, A. (1984) "A relaxation algorithm for generic chemical structure screening." *J. Chem. Inf. Comput. Sci.* 24: 235–241.

20. Chen, L. (2004). "Substructure and maximal common substructure searching." *Comput. Med. Chem. Drug Discovery.* pp. 483–513.

21. Prentis, R.A., Lis, Y. and Walker, S.R. (1988). "Pharmaceutical innovation by the seven UK-owned pharmaceutical companies (1964–1985)." *Br. J. Clin. Pharmacol.* 25: 387–396.

22. Pawel Baranczewski, Per Olof Edlund and Hans Postlind. (2006). "Characterization of the cytochrome P450 enzymes and enzyme kinetic parameters for metabolism of BVT.2938 using different

in vitro systems." *Journal of Pharmaceutical and Biomedical Analysis.* Vol. 40. 5: 1121–1130.

23. Singer, S. J. and Nicolson, G. L. (1972). "The Fluid mosaic model of the structure of cell membranes." *Science.* 175: 720–731.

24. Conradi, R.A., Burton, P.S. and Borchardt, R.T. (1996). "Physicochemical and biological factors that influence a drug's cellular permeability by passive diffusion methods." *Princ. Med. Chem.* 4: 233–252.

25. Viswanadhan, V.N., Reddy, M.R., Bacquet, R.J. and Erion, D.M. (1993). "Assessment of methods used for predicting lipophilicity: Application to nucleosides and nucleoside bases." *J. Comput. Chem.* 9: 1019–1026.

26. Klopman, G., Li, J.Y., Wang, S. and Dimayuga, M. (1994). "Computer automated logP calculations based on an extended group contribution approach." *J. Chem. Inf. Comput. Sci.* 34: 752–781.

27. Wang, R., Fu, Y. and Lai, L. (1997). "A new atom-additive method for calculating partition coefficients." *J. Chem. Inf. Comput. Sci.* 37: 615–621.

28. Palm, K., Stenberg, P., Luthman, K. and Artursson, P. (1997). "Polar molecular surface properties predict the intestinal absorption of drugs in humans." *Pharm. Res.* 14: 568–571.

29. Palm, K., Luthman, K., Ungell, A., Strandlund, G., Beigi, F., Lundahl, P. and Artursson, P. (1998). "Evaluation of dynamic polar molecular surface area as predictor of drug absorption:Comparison with other computational and experimental predictors." *J. Med. Chem.* 41: 5382–5392.

30. Clark, D.E. (1999). "Rapid calculation of polar molecular surface area and its application to the prediction of transport phenomena. 1. prediction of intestinal absorption." *J. Pharm. Sci.* 88: 807–814.

31. Kelder, J., Grootenhuis, P.D.J., Bayada, D.M., Delbressine, L.P.C. and Ploemen, J. (1999). "Polar molecular surface as a

dominating determinant for oral absorption and brain penetration of drugs." *Pharm. Res.* 16: 1514–1519.

32. Young, R.C., Mitchell, R.C., Brown, T. H., Ganellin, C.R., Griffith, R., Jones, M., Rana, K.K., Saunders, D., Smith, I.R., Sore, N.E. and Wilks, T.J. (1988). "Development of a new physicochemical model for brain penetration and its application to the design of centrally acting H2 receptor histamine antagonists." *J. Med. Chem.* 31: 656–671.

33. Seiler, P. (1974). "Interconversion of lipophilicities from hydrocarbon/water systems into the octanol/water system." *Eur. J. Med. Chem.* 9: 473–479.

34. Crivori, P., Cruciani, G., Carrupt, P.A. and Testa, B. (2000). "Predicting blood-brain barrier permeation from three-dimensional molecular structure." *J. Med. Chem.* 43, 11: 2204 –2216.

35. Herve, F., Urien, S., Albengres, E., Duche, J.C. and Tillement, J. (1994). "Drug binding in plasma. A summary of recent trends in the study of drug and hormone binding." *Clin. Pharmacokinet.* 26: 44–58.

36. Colmenarejo, G., Alvarez-Pedraglio, A. and Lavandera, J.L. (2001). "Cheminformatic models to predict binding affinities to human serum albumin." *J. Med. Chem.*, 44: 4370–4378.

37. Carter, D.C. and He, X.M. (1990). "Structure of human serum albumin." *Science.* 249: 302–303.

38. Boelsterli, U.A. (2002). "Mechanisms of NSAID-induced hepatotoxicity: Focus on nimesulide." *Drug Safety.* 25(9):633–648.

39. Balasubramaniam, J. "Nimesulide and neonatal renal failure." *The Lancet.* Vol. 355. 9203: p. 575.

40. Keseruu, G. M. and Molnar, L. (2002). "METAPRINT: A metabolic fingerprint. application to cassette design for high-throughput ADME screening." *J. Chem. Inf. Comput. Sci.* 42: 437–444.

Bioclipse: an open source workbench for chemo- and bioinformatics

Open Source/Free Software

- Blue Obelisk-http://wiki.cubic.uni-koeln.de/dokuwiki/doku.php

- InChI - http://www.iupac.org/inchi/

- JMOL – http://jmol.sourceforge.net

- FROWNS - http://frowns.sourceforge.net/

- OpenBabel - http://openbabel.sourceforge.net/

- CML - http://cml.sourceforge.net/

- CDK - http://almost.cubic.uni-koeln.de/cdk/

- MMTK - http://starship.python.net/crew/hinsen/MMTK/

- http://www.bioclipse.net/

Journals

Journal of Chemical Information & Computer Sciences (ACS)

Journal of Molecular Graphics and Modelling (Elsevier).

Example of Free Online sADME/tox Tools

http://www.molinspiration.com/

http://www.molsoft.com/

http://www.chemaxon.com/

http://zinc.docking.org

http://www.syrres.com/

http://www.logp.com/

http://146.107.217.178/lab/alogps/

Some Online Compound Collections

http://chembank.med.harvard.edu

http://www.cermn.unicaen.fr/chimiothequehttp://www.genome.ad.jp/dbget/ligand.html

http://Ligand.info

http://zinc.docking.org

http://bioserv.rpbs.jussieu.fr/FAFDrugs.html Free

http://bioweb.ucr.edu/ChemMine

http://www.mdli.com

http://www.chemnavigator.com

http://www.ebi.ac.uk/chebi/ Dictionary of small molecules

http://www.bindingdb.org/

http://www.pdbbind.org/

http://kibank.iis.u-tokyo.ac.jp/

Recommended Books for Reading

1. *Chemoinformatics: A Textbook*, Edited by Johann Gasteiger and Thomas Engel, Wiley (2003). ISBN: 978-3-527-30681-7

2. *Chemoinformatics: Concepts, Methods, and Tools for Drug Discovery*, Edited by Jürgen Bajorath (Methods in molecular biology; v. 275), Humana Press (2007). ISBN 1-58829-261-4

3. *An Introduction to Chemoinformatics*, by Andrew R. Leach, Valerie, J. Gillet by Springer (2007). ISBN:1402062915

4. *Chemoinformatics: Theory, Practice, & Products*, Bunin, B.A., Siesel, B., Morales, G., Bajorath, J. Springer (2007). ISBN: 978-1-4020-5000-8

Websites of Related Interest

1. http://www.cheminformatics.org/

2. http://www-jmg.ch.cam.ac.uk/CIL/

3. http://cactvs.cit.nih.gov/

Journals

1. *Journal of Chemical Theory and Computation* (ACS)

2. *Journal of Molecular Structure: THEOCHEM* (Elsevier)

3. *Tetrahedron Computer Methodology* (Elsevier)

REVIEW QUESTIONS

1. Define chemoinformatics and elaborate Lipinski's rule of five.

2. Briefly explain the line notations used in chemoinformatics.

3. What are the types of databases? Enumerate the use of Chemical Abstract Service Registry database.

4. Search the PUBCHEM for naphthalene and report the number of hits for its substructures.

5. Write notes on inductive and deductive learning methods.

6. Explain in detail, the importance of ADMET properties in drug discovery.

5

QUANTITATIVE STRUCTURE ACTIVITY RELATIONSHIP (QSAR)

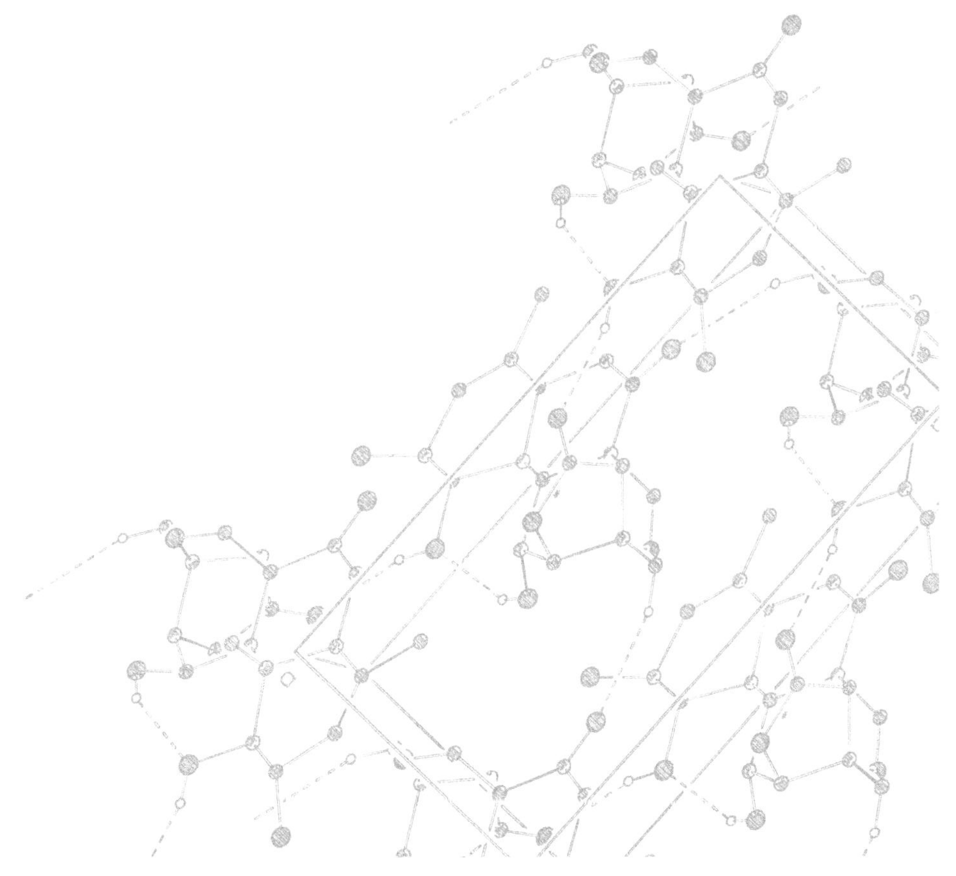

INTRODUCTION

Structure-based drug design,[1-4] has been a very successful approach in designing new/novel drugs. A speculative reason why there are loopholes in receptor modelling, which act as barriers in their complete success, is that too many parameters need to be estimated. This includes absence of consideration of conformational flexibility in the receptor protein, solvent dependency (dielectric constant), pH at the active site (protonated ligands or not), water-improved receptor binding,[5] induced fit[6,7] and many more features. Also factors like distortion energy of the ligand and its binding site, molecular electrostatic potential and dielectric constant at the binding site, dipole moment of the ligand and local dipole moment at the binding site, binding enthalpy of the ligand–protein complex, repulsive effects (e.g. $-O...O^-$), solvation enthalpy and entropy of the complex — all these need to be considered. In other words, it has not been possible to exactly mimic the environment in which the protein and the drug interact in the human system.

Quantitative structure activity relationship (QSAR) is the name usually applied to methods that correlate molecular structure of the ligands/drugs to some kind of *in vitro* or *in vivo* biological property. It can be classified as ligand-based drug design approach in the sense it is applicable, when the structure of the target/receptor is known or not.

QSAR studies can show way in identifying the features that makes a molecule active or inactive. In terms of the lock and key metaphor, since 'nature of the lock' is not known, it is not possible to judge which keys work and which do not (since no docking can be done without a receptor!). QSAR can be considered as the method of trying to build a model to explain why some keys work and others do not. It attempts to link activity data with descriptors chosen via identification of the "rules" that can be further used to guide chemical synthesis when new chemical entities are developed. The aim of such investigations is to produce a suitably robust model (which can be an equation or map) capable of reliable predictions for novel chemical species.

When applied to modelling of physico-chemical properties of molecules (boiling point, vapour pressure, etc.) it is called Quantitative Structure–Property Relationship (QSPR).

QSAR can predict quantities such as the binding affinity, pharmacokinetic parameters, toxicity and some environmental related parameters of a given molecule. QSAR-based (*in silico*) analysis may be better regarded as an exercise to screen or filter drug candidates, before they are subjected to more intensive calculations such as docking or an experimental measurement of activity (*in vitro*) and finally under real conditions (*in vivo*). Typically, this step will pick up a dozen of drug candidates from a library of millions of well-studied molecules.

Utility of the QSAR approach includes prediction of property/ activity, refinement of synthetic ligand (it can suggest as to what type of, say, a side chain modification will improve or decrease the bioactivity of a compound), etc.

The input for such a study are the structures of molecules (known to possess certain biological activity) and their bioactivity profile. The activity profile needs to be generated by experimental work and most importantly should be in quantitative terms. A minimum of twenty compounds, known as the training set, with varying levels of biological activity, is needed to derive a valid QSAR equation.

THE EVOLUTION OF QSAR

Richardson[8] in 1868 showed that the toxicities of ethers and alcohols were inversely related to their water solubility. Richet[9] (1893) demonstrated a relationship between the narcotic activity of alcohols and their molecular weight and Overton[10] (1897) and Meyer[11] (1899) independently showed a relationship between the anaesthetic action of many compounds and their lipophilicity represented by olive oil–water partitioning coefficients; but the main impact on the development of the concepts of QSAR was from the work of Hammett[12] (1937) and Taft[13] (1952) which discussed the effect of functional group substitution/replacement on organic reactions.

These contributions constitute together, the mechanistic basis for the development of the QSAR paradigm. Corwin Hansch's research publications[14,15] in the 1960s about the relationship of plant growth regulators and their dependency on Hammett constant and hydrophobicity laid the foundation of QSARs. The Hansch analysis introduced a way to predict biological activity from theoretically derived molecular descriptors. Hansch and co-workers investigated the possibility of expressing a relationship between structural and physico-chemical properties and biological activity, quantitatively.

Some of the descriptors that evolved out of this study are listed in Table 5.1. In the Hansch analysis, the biological activity log (1/C) was correlated with the descriptors using multiple linear regression. The Hansch equation is given as

$$\log\,(1/C) = -a(\log P)^2 + b\log P + c\sigma + d\,Ea + e \tag{5.1}$$

where *a*, *b*, *c*, *d* and *e* are constants obtained through statistical analysis of the data.

A typical QSAR would look like equation (5.1) (derived from the training set) from which the constants would be calculated. Since the molecular descriptors (log P), σ, *Ea* are dependent only on the structure, calculating these descriptors (*in silico*) and substituting in the equation would give the activity of a compound, whose biological activity has not been tested; but the accuracy of the prediction depends on the accuracy of the biological data of the training set.

Table 5.1   Some of descriptors used in QSAR studies

Descriptor	Symbol	Type
Formula weight	FW	Steric
Hammett constants	σ_m, σ_p	Electronic
Tafts polar constant	σ^*	Electronic

(Contd.)

Table 5.1 (Continued)

Descriptor	Symbol	Type
Tafts steric parameter	E_s	Electronic
Hansch aromatic fragment	Π	Steric
Lipophilicity	log P	Lipophilic
Lipophilicity (pH = 7.4)	log D	Lipophilic
Connectivity index	$^1\chi$	Lipophilic
Molar refractivity	MR	Steric
Verloop Sterimol	L,B	Steric
Eudismic index	EI	Electronic
Ionization constant	pK_a	
van der Waals volume	VdWV	Steric
van der Waals area	VdWA	Steric
Connolly surface volume	CoVo	Steric
Connolly surface area	CoAr	Steric
Electronic energy	ELEC	Electronic
Core–core interaction	CoCo	Steric
Heat of formation	HoFo	Electronic
Highest occupied molecular orbital	HOMO	Electronic
least unoccupied molecular orbital	LUMO	Electronic
Dipole moment	Dipo	Electronic
Point charges	chaX	Electronic

The ideas behind the Hansch analysis are as valid today as in 1962 since theoretically (not necessarily) generated descriptors are assumed to be independent of the ligands' conformations. More important, if the predictive ability of the model is high enough, valuable time will not be spent synthesizing inactive compounds.

Since then, with the increasing knowledge in chemistry and the exploding power of computers, more and more QSAR models have been developed. QSARs development has now become an important branch of **chemometrics**, the science of the application of mathematical or statistical methods to chemical data.

The most important of these are steric descriptors (related to size and shape of the chemical), electronic descriptors (related to the ability of the chemical to undergo reactions) and the polarity descriptors (water-immiscible; related to transfer across cell membranes).

Some typical QSAR applications include

1. *Calculation of single molecular species properties* like boiling point, vapour pressure, critical temperature, flash point, melting point, refractive index, density, viscosity.

2. *Interactions between different molecular species* like octanol–water partition coefficient, aqueous solubility of liquids and solids, etc.

3. *Biological properties* such as sweetness correlations, mutagenicity, general toxicities, algistatic activity, etc.

All QSAR techniques assume that

1. The compounds interact with the same biological target in a non-covalent manner (hydrogen bonds, hydrophobic, hydrophilic interaction)

2. The interaction of similar structured compounds are similar with the functionalities in the binding site

3. This assumes a rigidity of the interacting systems with no molecular dynamics into consideration

The best QSAR relationship will be obtained with:

◈ Closely related active molecules, all having the same mode of action

◈ A good spread of activity in the test set to improve statistical validity

◈ An assay which is close to the site of action (e.g. binding or enzyme assay)

The biological activity is usually expressed as (1/C) where, C is the desired concentration of the drug to achieve a defined level of activity; inverse because more potent ones achieve the level at much lower concentrations or IC_{50} (Drug concentration that is required for 50% inhibition *in vitro*) or EC_{50} (Plasma concentration required in obtaining 50% of a maximum effect *in vivo*), depending on the type of assay.

Descriptors are generally selected in one of two ways. Firstly, it can be a presumption search, wherein descriptors which might yield the best model for the property in question is chosen on a criterion. Secondly, a large number of descriptors (perhaps several hundred) are generated and a statistical method such as step-wise regression/principle component analysis/genetic algorithm/ method of partial least squares is applied to select the 'best' descriptors, i.e., those that give the best correlation with the property in question.

The aim and utility of QSAR studies are to

◈ Develop quantitative structure–activity relationships

◈ Predict the properties and activities of untested molecules

◈ Optimize the properties of a lead compound

◈ Compare different QSAR models statistically and visually

◈ Generate hypotheses about the characteristics of a receptor-binding site

◈ Validate models of receptor-binding sites

◈ Prioritize compounds for synthesis or screening

CASE STUDY

The neuronal nicotinic acetylcholine receptors are transmembrane proteins (ion channel modulators) that can exist under different conformations, at least one forming a pore through the membrane connecting the two neighbour compartments. The equilibrium between the various conformations is affected by the binding of ligands on the channels. Basically, the ligands "open" or "close" the channel.

$$K_i = -141.363 + 1449.55\,(SIC)^2 + 0.000602\,(\text{Jurs-}WNSA\text{-}2)^2 - 47.5526\,(IC)^2 + 2.43134\,(\log Z) \qquad (5.2)$$

The information-theoretic topological indices[16] are found in equation (5.2). IC (Informational content) and SIC (structural informational content) are present and may be identifying steric qualities of the molecules and log Z (logarithm of the Hosoya index), which deals with connectivity of the molecule. It also appears that the Jurs descriptors are effective in describing the SAR of the quaternary ammonium salts. The GFA selected a surface-weighted, partial negative charge descriptor, Jurs-WNSA, which is most likely related to a portion of the molecule, possibly the aromatic pyridinium ring.

Table 5.2 gives a list of some of the descriptors used in QSAR studies.

Table 5.2 Types of descriptors in QSAR studies

Constitutional descriptors	Topological descriptors	Charge descriptors	Quantum chemical descriptors	Walk and path counts	Geometric descriptors
Molecular weight, counts of atoms and bonds, counts of rings, etc.	Wiener index, Kier and Hall indices , information indices, total structure connectivity index, Pogliani index, ramification index, polarity number, average vertex distance degree, mean square distance index (Balaban), Schultz Molecular Topological Index (MTI), square reciprocal distance sum index, quasi-Wiener index (Kirchhoff number),	Maximum positive charge, maximum negative charge, total positive charge, total negative charge, total absolute charge (electronic charge index—ECI), mean absolute charge (charge polarization), total squared charge, relative positive charge, relative negative charge, submolecular polarity parameter, topological electronic descriptor,	Dipole moment components, polarizability, sigma and pi-bond orders, molecular orbital energies, frontier molecular orbital reactivity indices, normal modes, energy partitioning terms, thermodynamic properties, electrostatic surface descriptors	Molecular walk counts, total walk count, self-returning walk counts, molecular path counts, molecular multiple path counts, total path count, conventional bond-order ID number, Randic ID number, Balaban ID number, ratio of multiple path count over path count, difference between multiple path count and path count	Principal moments of inertia, molecular volume, total solvent-accessible surface, molecular shadow projections

spanning tree number, hyper distance-path index, reciprocal hyper-distance-path index, detour index, hyper-detour index, reciprocal hyper-detour index, distance/detour index, all-path Wiener index, Wiener-type index from Z weighted distance matrix (Barysz matrix), molecular electro-topological variation, E-state topological parameter, Kier symmetry index eccentricity, mean distance degree deviation, unipolarity, centralization, variation.	topological electronic descriptor (bond-restricted), partial charge weighted topological electronic descriptor, local dipole index.			

HYBRID DESCRIPTORS

Hybrid descriptors are generally combinations of electronic or topological descriptors and geometric descriptors and in general characterize the distribution of a molecular feature over the whole molecule. Examples include the Charged Partial Surface Area (CPSA),[17] Hydrophobic Surface Area (HPSA)[18] and hydrogen bonding[19,20] descriptors. The importance of these descriptors is that they provide localized information regarding molecular features. Thus, in the case of the HPSA descriptors, it is possible to obtain specific values of the hydrophobicity for different regions of the molecule as well as a global value for the whole molecule. Examples include the atomic constant weighted hydrophobic and hydrophilic surface areas, total hydrophobic constant weighted hydrophobic surface area and the relative hydrophobicity. The functional forms of these descriptors are given below.

$$PPHS - 2 = \Sigma(+SA_i)(+\log P_i)$$
$$PNHS - 2 = \Sigma(-SA_i)(-\log P_i)$$
$$THWS = \Sigma(SA_i)(\log P_i)$$

$$RPH\text{-}1 = \frac{\text{Most hydrophobic atom constant}}{\Sigma \log P_i}$$

where SA_i is the surface area for the ith atom and $\log P_i$ is the hydrophobic constant for the ith atom and the + and − symbols indicate a hydrophobic or hydrophilic atom respectively.

The CPSA descriptors (Table 5.3) are similar in concept to the HPSA descriptors. In this case, the surface area values are combined with partial charges leading to 25 descriptors. In addition, a number of CPSA descriptors specific to certain atoms (such as N and O) are also calculated. These descriptors are similar in concept to the polar surface area (PSA) descriptors[21, 22] which have been shown to be very useful in studies of intestinal absorption[23] and blood-brain barrier crossing.[24] The development of the topological polar surface area (TPSA)[25] allows the rapid evaluation of polar surface areas using only connectivity information (SMILES strings) and a library of fragment contributions. By combining molecular surfaces with atomic properties, these descriptors are useful both in 2D as well as 3D QSAR methods. In addition, surface–property descriptor types usually have simple physical interpretations and have been shown to be quite information-rich.[26, 27]

Table 5.3 Charged partial surface area descriptors

Descriptor	Descriptor label	Formula
Partial positive surface area	PPSA-1	$\Sigma\,(+SA_i)$
Partial negative surface area	PNSA-1	$(\Sigma\,(-SA_i)$
Total charge weighted PPSA	PPSA-2	$(\Sigma\,(+SA_i))\,(Q_T^+)$
Total charge weighted PNSA	PNSA-2	$(\Sigma\,(-SA_i))\,(Q_T^-)$
Atomic charge weighted PPSA	PPSA-3	$(\Sigma\,(+SA_i))\,(Q_i^+)$
Atomic charge weighted PNSA	PNSA-3	$(\Sigma\,(-SA_i))\,(Q_i^-)$
Difference in charged partial surface areas	DPSA-1 DPSA-2 DPSA-3	$[(PPSA\text{-}1) - (PNSA\text{-}1)]$ $[(PPSA\text{-}2) - (PNSA\text{-}2)]$ $[(PPSA\text{-}3) - (PNSA\text{-}3)]$
Fractional charged partial surface areas	FPSA-1 FNSA-1 FPSA-2 FNSA-2 FPSA-3 FNSA-3	(Charged partial surface area)/(total molecular surface area)
Surface weighted charged partial surface areas	WPSA-1 WNSA-1 WPSA-2 WNSA-2 WPSA-3 WNSA-3	$(CPSA)$ (Total mol. Surf. area)/1000
Relative positive charge	RPCG	(Charge of most positive atom)/ (sum total positive charge)

(Contd.)

Table 5.3 (Continued)

Descriptor	Descriptor label	Formula
Relative negative charge	RNCG	(Charge of most negative atom)/ (sum total negative charge)
Relative positive charged surface area	RPCS	(SAmpos) (RPCG)
Relative negative charged surface area	RNCS	(SAmneg)(RNCS)

* $(+SA_i)$ and $(-SA_i)$ are the surface area contributions of the 'i'th positive or negative atom in the molecule. Q_i^+ and Q_i^- are the partial atomic charges for the 'i'th positive and negative atoms, while Q_T^+ and Q_T^- are the sum total positive and negative charges for the molecule.

QSAR by Eigen Value Analysis (EVA) uses a normal mode calculation to simulate the infrared spectrum of a molecule, with each descriptor representing the intensity of the spectrum for a small range of frequencies.[28] The EVA descriptor sets for each molecule are submitted to QSAR analysis after a number of preprocessing steps.

THE DERIVATION OF A DEPENDABLE MODEL IN QSAR STUDIES

The best way for determining the predictive ability of a QSAR model is to predict the property in question for a number of compounds that were not used in the training set, but for which the measured value of the property is known; such a set of compounds is called a **test set**. The test set compounds must be reasonably similar to those of the training set; that is, they must lie within the applicability domain of the QSAR. This is often achieved by dividing the total number of compounds into two groups; the larger group forms the training set and the smaller group (typically 5–50% of the total) forms the test set. If the standard error for the test set is much larger than that for the training set, then the QSAR does not have good predictivity and it should not be used for predictive purposes.

If the total number of compounds is small, then it may not be practicable to split it into training and test sets. In that case, a procedure called internal cross-validation can be used, whereby each compound in turn is deleted from the training set, the QSAR is developed with the remaining compounds and is used to predict the property value of the omitted compound. That compound is then returned to the training set and a second compound is deleted and so on until every compound has been left out in turn. A cross-validated r^2 value is then calculated which is an indicator of the internal predictivity of the QSAR. The most popular techniques for quantitative endpoints are simple or multiple linear regression,[28, 29] principal components regression, partial least squares regression,[30] as well as regression trees and neural networks.

If the activity property is qualitative, discriminent analysis, decision trees or distance-based similarity analysis is often applied to model its relationship with the physiochemical explanatory variables.

STEPS INVOLVED IN A QSAR STUDY

1. The descriptors are calculated.

2. The best descriptors from a larger set of relevant descriptors are identified.

3. Molecular descriptors are mapped into the biological activity, wherein the relationships can be often complex, or linear.

4. The model is validated to determine its predictive ability and to generalize to new molecules that are not in the training set.

STATISTICAL METHODS

Some of the statistical methods used in QSAR analysis are briefly discussed below.

Data reduction techniques such as **Principal Component Analysis (PCA)** overcome the requirement of a high observation-to-parameter ratio. By reducing the number of variables that describe biological activity or chemical properties to a fewer number of independent orthogonal components, regression can be performed on these principal components. The result is that redundancies are removed and inter-correlated data is minimized.

Partial least squares (PLS) goes one step further than PCA by including cross-validation, a technique of leaving out components to be predicted by the relationship established by the other compounds. The actual predictive ability of the final model is then evaluated by how well it predicts the unprocessed, unbiased data. Although PCA and PLS can produce highly predictive QSAR models, their main drawbacks lie in their limited ability to derive interpretable models.

Stepwise regression methods, such as forward-stepping linear regression, have also been of much utility because they can produce models with a reasonable level of interpretability and are easily applied to original descriptor sets. These methods, however, rely on obtaining sufficient response levels from individual variables in isolation. With extremely large data sets, the signal-to-noise ratio of a single variable is not always apparent.

Genetic function approximation (GFA) is part of a powerful class of computational techniques known as genetic algorithms. Incorporated into QSAR model development, genetic algorithms help in finding optimum solutions for combinatorial problems. Genetic algorithms offer a significant advantage in that, unlike other methods, they consider variables in combination with one another instead of just in isolation. Genetic algorithms also maintain the use of original descriptors without converting the descriptors into principal components, thereby retaining the desired level of interpretability of the final QSAR model.

Simulated annealing searches the descriptor space for optimal subsets one string at a time. It begins with an initially random string of descriptors and replaces one or more of the descriptors with new descriptors from the reduced pool. Each new subset is evaluated by an appropriate cost function—typically an error minimization.

If the cost function of the new subset is better than the previous subset, then the new subset is appropriately stored in the list of best models. If the cost function is worse, then a probability function is used to determine if the algorithm should take a detrimental step, that is, proceed with a mutation of the new subset or revert back to the previous subset of descriptors and attempt a new mutation. The ability to take many more detrimental steps early in the optimization reduces the risk of converging in a local error to a minimum. Thus as the optimization proceeds, detrimental steps become more difficult to take.

TYPES OF QSAR METHODS

QSAR methodologies can be broadly divided into three groups. First, 2D methodologies do not consider the 3D structure of a molecule directly. Instead, the molecule is represented by a set of molecular descriptors, numerical values characterizing various aspects of molecular structure. Together with the observed activity, a predictive model is built. It should be noted that even though some descriptors are based on 3D coordinates, the method as a whole considers only the observed property and the descriptors and hence is 2D in nature. Two-dimensional approach has a number of advantages. First, owing to the variety of molecular descriptors available, optimized coordinates are not always required. In fact, connectivity information (in the form of SMILES strings or an adjacency matrix) alone can be used to develop QSAR models. As a result, models using these types of descriptors (termed topological descriptors) can be built rapidly for very large sets of molecules. However, these types of descriptors are in general quite abstract and so if the model is to be analysed to extract information regarding structure–property trends, other, more physically meaningful descriptors will generally be required. Second, this approach avoids the alignment step and thus can be used in the absence of experimental information regarding the binding of a molecule to its target. The downside to the 2D QSAR methodology is that it does not provide a detailed answer to a number of questions regarding a molecule's activity. That is, by representing structural information in the form of descriptors, the aspects of a molecule's activity such as its absorption or degradability properties are hidden by a layer of abstraction or not addressed at all. Thus a molecule might be observed to have low activity. A 2D model may not be able to indicate whether this is due to its inability to bind to the target or whether this is due to its inability to cross the cell membrane. The point is that, in a 2D QSAR model, a lot of information about various aspects of a molecule's activity are combined together and are not always individually apparent. Though interpretation methods for linear QSAR models exist, they are obviously restricted to the information encoded by the descriptors in the model. This means that though 2D QSAR models are certainly very useful, especially for screening purposes, they should be used in conjunction with other types of models to fully understand the role that various structural features play in determining the activity of a molecule.[31]

The second type of methodology is 3D in nature and is exemplified by the Comparative Molecular Field Analysis (CoMFA)[32] approach.

Molecular fields basically are three-dimensional representations of the steric, electrostatic and hydrophobic surroundings of a molecule. A molecular field is generated by enclosing the molecule in a three-dimensional grid (Figure 5.1) and assigning non-bonded interactions between a probe atom and the molecule in each grid point. Obviously, the difference between two different types of fields is the algorithm with which the non-bonded interactions are calculated.

It is of importance because they allow for the simulation of directional forces: hydrogen bonds, metal–ligand contacts, polarization effects, and interaction between electric dipoles. Such forces are known to play a key role for both molecular recognition and selective ligand binding.

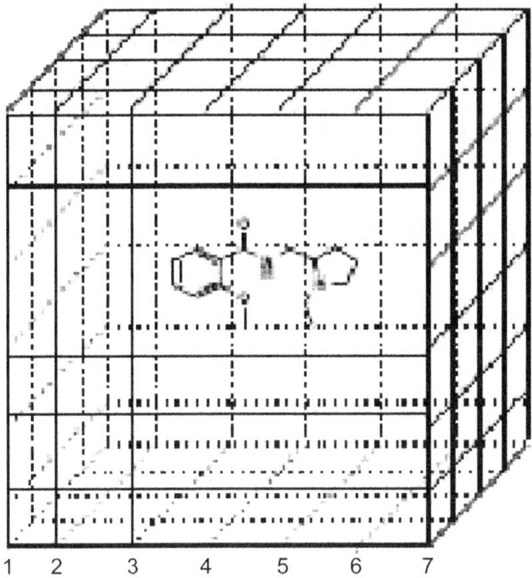

Figure 5.1 The three-dimensional grid used in CoMFA to generate molecular field descriptors. For clarity, grid points within the grid are omitted.

In SYBYL/CoMFA the steric field interaction energies (E_{ste}) are Lennard–Jones potentials, also referred to as steric 6-12 potentials,[33] which are sensitive to changes in the distance between the probe and the atoms (r_i) as can be seen in equation 5.3. N is the number of atoms in the molecule; A and B are constants characteristic for the probe atom type and the type of the ith atom in the molecule, respectively.

$$E_{ste} = \sum_{i-1}^{N}\left[\frac{A}{r_i^{12}} - \frac{B}{r_i^{6}}\right] \tag{5.3}$$

The electrostatic field interaction energies are less influenced by the distances between the probe and the atoms but instead the charge of the probe and the point charges of the atoms are important. In addition, E_{ele} is very sensitive to spatial dielectric behaviour of the environment[34] and a distance-dependent dielectric term has been proposed.[35] The magnitude of the electrostatic potential (E_{ele}) between two ions with charges Q and q separated by a distance r is given by Coulomb's law

$$E_{ele} = \sum_{i=1}^{N} \frac{Qq_i}{K\zeta} \left[\frac{1}{r} + \frac{(\zeta - \varepsilon)/(\zeta + \varepsilon)}{\sqrt{r_2 + 4s_Q s_q}} \right]$$

where,

> N is the number of atoms in the molecule,
>
> Q is the charge of the probe atom,
>
> q_i is the point charge on the ith atom and
>
> K is a constant term.

In this formula, a homogenous protein phase and a homogenous solution phase with dielectrics ζ and ε, respectively, are assumed to be present. The depth of each protein atom (S_p) in the protein phase is assessed by counting the number of neighbouring protein atoms whose nuclei lie within 4 Å. For the probe atom the depth (S_Q) is calculated similarly. Consequently, the above equation, leads to an effective dielectric of ζ when the pairwise groups of atoms are so deep in the protein and so close together that the solvent effects can be neglected. However, when one or both of the atoms approach the surface of the protein the effective dielectric becomes ($\zeta + \varepsilon$)/2 since the term $4S_Q S_q$ is set to zero.

In analogy with the Hansch analysis, in CoMFA the molecular fields are correlated with the biological activity. CoMFA takes the three-dimensional conformations of the molecules into consideration, hence, several crucial items need to be considered.

First, in order to find low-energy conformations, or rather the global minimum energy conformation of each compound under investigation, conformational analyses is conducted.

According to the Boltzman distribution, a low-energy conformation is more abundant than a high-energy conformation and consequently, also more likely to be involved in the ligand–receptor interaction.

Second, all molecules must be aligned in the same coordinate system and several options to perform this are possible. If a pharmacophore is available, one might choose to superimpose all molecules on mutual and likely interaction points with the receptor. The compounds could be docked into the active site of a receptor homology model or, alternatively, the molecular fields could be superimposed in a least square manner.

Independent of whichever alignment approach that is employed, in CoMFA the differences between the aligned molecular fields are correlated with the biological activity. Therefore, an alignment procedure where the global overlap between structurally related compounds are maximized, is likely to perform just as good as a more elaborated and rational alignment . This is the case when the ligands are flexible and the rationale is purely statistical. If the alignment is not performed with maximized overlap between the molecules, an increased level of insignificant variation, i.e., noise, is inevitable. Noise may or may not affect the predictability detrimentally.

Third, molecular fields are generated first when the molecules are properly aligned.

The CoMFA studies result in contour plots—steric and electrostatic.

In steric contour plots

- ◇ **Green** contours indicate regions where an increase in steric bulk will enhance activity

- ◇ **Yellow** contours indicate regions where an increase in steric bulk will reduce activity

In electrostatic contour plots

- ◇ **Blue** contours and **red** contours correspond to region where an increase in positive or negative charge, respectively will enhance activity.

This type of output would give a direction to the synthetic chemists as to what type of modifications in the basic structure of a molecule would result in enhancing the activity or otherwise.

CASE STUDY

4-substituted phenyl 2,6-dimethyl 3,5-bis-N-(substituted phenyl carbamoyl- 1,4-dihydropyridines have been reported to be effective against *M. tuberculosis* H37Rv.[36] The basic skeleton of 1,4-dihydropyridines used in the training set at test set are given in Figure 5.2 and Figure 5.4 respectively. To further explore the structural requirements of 1,4-dihydropyridines for the antitubercular activity, two methods of three-dimensional quantitative–structure activity relationship (3D QSAR), comparative molecular field analysis (CoMFA) and comparative molecular similarity indices analysis (CoMSIA) were performed.[37] Table 5.4 gives the bioactivity profile of the molecules used in the training set.

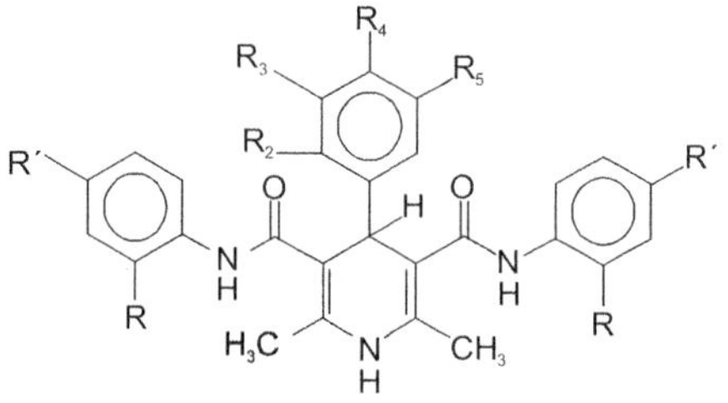

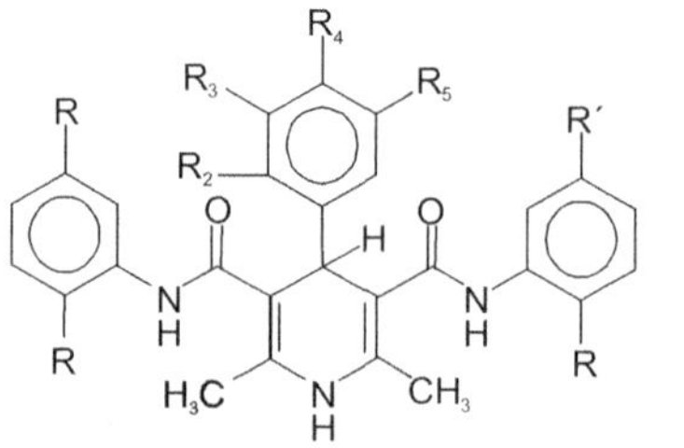

Figure 5.2 Structures of molecules in the training set

Table 5.4 The experimental bioactivity of the molecules in the training set

Compound	R	R′	R_2	R_3	R_4	R_5	Observed activity
1	H	Cl	H	NO_2	H	H	– 0.3888
2	H	Cl	OH	H	OCH_3	H	0.6297
3	H	Cl	H	OCH_3	OCH_3	OCH_3	1.1949
4	H	NO_2	Cl	H	Cl	H	1.3802
5	H	NO_2	H	H	$N(CH_3)_2$	H	1.5096
6	H	NO_2	H	H	Cl	H	0.2126
7	Cl	H	H	H	SCH_3	H	0.5496
8	Cl	H	Cl	H	H	H	–0.6585
9	Cl	H	H	H	Cl	H	–0.6886
10	Cl	H	Cl	H	Cl	H	–0.9079
11	Cl	H	SH	H	H	H	–0.9542
12	OCH_3	H	H	H	SCH_3	H	1.1233
13	OCH_3	H	Cl	H	Cl	H	1.0047
14	OCH_3	H	H	NO_2	H	H	0.2498
15	OCH_3	H	Cl	H	H	H	–0.2311
16	H	H	H	NO_2	H	H	0.4775
17	CH_3	CH_3	H	H	SCH_3	H	0.1580
18	CH_3	CH_3	H	H	H	H	–0.8653
19	CH_3	CH_3	H	H	OCH_3	H	–0.1580
20	H	NO_2	Cl	H	H	H	1.5096
21	H	NO_2	Cl	H	Cl	H	1.6901
22	H	NO_2	H	H	H	H	1.1949

The molecules in the training set are first superimposed to get a general conformation which is supposed to be the bioactive conformation.

Using the above structures and data, 3D QSAR analysis was carried out and the result was applied to a test set of compounds (Figure 5.3, Table 5.5).

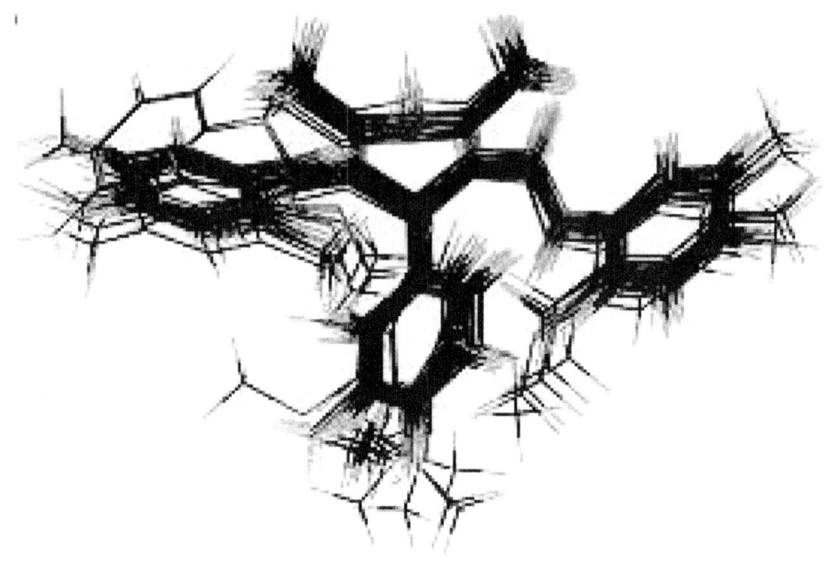

Figure 5.3 Superposition of derivatives of 1,4-dihydropyridines in the training set

Table 5.5 The variation in the structure (23–33), the observed activity along with the activity predicted by CoMFA and CoMSIA

								Predicted activity	
Compound	R	R$'$	R$''$	R$_2$	R$_3$	R$_4$	Observed activity	CoMFA	CoMSIA
23	H	H	NO$_2$	H	H	SCH$_3$	1.6901	2.1747	2.2078
24	H	NO$_2$	H	H	H	SCH$_3$	1.5096	1.2541	1.3834
25	H	NO$_2$	H	Cl	H	H	1.2787	0.8733	0.8437
26	OCH$_3$	H	H	H	H	N(CH$_3$)$_2$	1.1233	1.3892	1.4809
27	H	H	NO$_2$	OH	H	H	1.0606	1.2600	1.4481
28	H	Cl	H	H	H	OH	0.5754	−0.1829	−0.3523
29	H	H	H	H	H	SCH$_3$	0.3474	0.7904	0.2145
30	CH$_3$	CH$_3$	H	H	NO$_2$	H	0.0521	−0.7104	−0.4597
31	Cl	H	H	H	H	OCH$_3$	−0.3679	−0.1816	0.0485
32	CH$_3$	H	CH$_3$	OH	H	H	−0.4319	−0.8590	−1.0642
33	Cl	H	H	OH	H	H	−1.1949	−0.5632	−0.6004

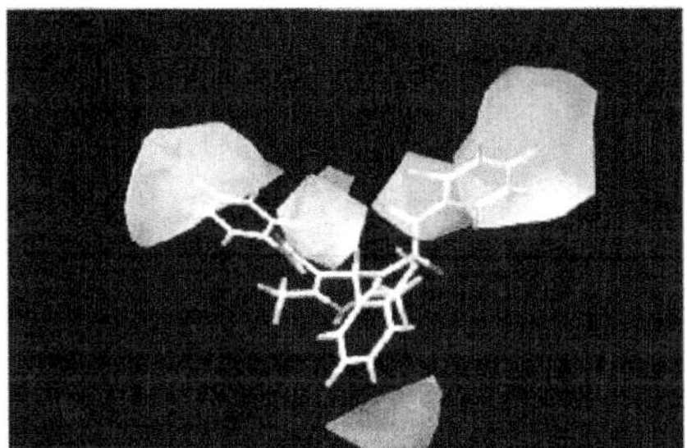

Figure 5.4 The basic structure of molecules in the test set

The resultant QSAR equations depicting the relation between the anti-tubercular activity and the molecular discriptors are given in Table 5.6.

Lack-of-fit (LOF) is a measure of the fitness of each equation. The equation with the least value of LOF and $r^2 \sim 1$ are considered robust. The contour maps as the resultant of the studies are given in Figure 5.5 and Figure 5.6.

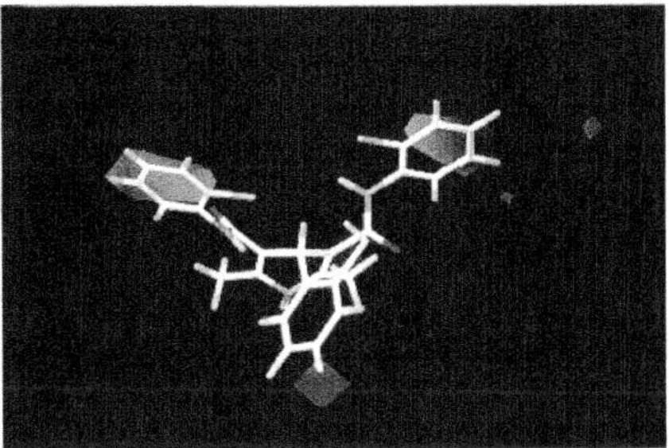

Figure 5.5 Compound **11** is shown inside the field. Green contours (contribution level of 80%) represent areas where steric bulk will enhance activity and yellow contours (contribution level of 20%) highlight areas, which should be kept unoccupied for increased activity. (See Plate 5)

Figure 5.6 Compound **11** is shown inside the field. Blue contours (contribution level of 80%) represent regions where an increase of positive charge will enhance activity and red contours (contribution level of 20%) highlight areas where more negative charge is favoured. (See Plate 6)

Table 5.6 QSAR equations derived using genetic function algorithm

No	Eq	LOF	r^2	F value	r^2_{cv}	r^2_{pred}
1	Activity = 6.2764 – 10.5279 (Density) + 0.4485 (HbondAcc) + 0.004036 (PMI_X)	0.310	0.777	20.924	0.657	0.424
2	Activity = –5.3487 + 0.5432 (RotlBonds) + 0.02970 (Foct) + 0.003497 (PMI_X)	0.321	0.769	19.986	0.657	0.551
3	Activity = –5.6129 + 0.03111 (Fh2o) + 0.5679 (RotlBonds) + 0.003343 (PMI_X)	0.323	0.768	19.824	0.654	0.549
4	Activity = –4.2602 +0.2817 (HbondAcc) + 0.3468 (RotlBonds) –0.1202 (Dipole_mag)	0.333	0.761	19.095	0.677	0.544
5	Activity = –5.05404 – 0.1103 (Dipole_Z) + 9.6037 (FPSA_P) + 0.008149 (PMI_X)	0.223	0.840	31.456	0.781	0.061
6	Activity = –5.7508 + 5.2509 (FPS_PA) + 0.003547 (PMI_X) + 0.4268 (RotlBonds)	0.284	0.796	23.444	0.691	0.560

PITFALLS OF MOLECULAR FIELD ANALYSIS

◇ The quality of the Lennard–Jones energy calculation and the Coulomb energy calculation depends heavily on the force field and charges

◇ The alignment or fit of the analogues is the most questionable part

◈ The fact that only one conformation can be subjected to the procedure is another weakness, which, however, can be circumvented by several techniques.

A pitfall that has not yet been widely recognized occurs when the fit of the compounds is not performed with the "rigid body" assumptions but with fit constraints. The difference in energy between the conformer after the fit and its minimal energy counterpart should be added to the measure of activity. Since an energy difference of only 4 kcal/mole corresponds to a 1000-fold decrease in activity; this pitfall deserves more awareness.

◈ The density and spacing of the grid is another determining factor.

◈ Statistically, the number of compounds omitted in the cross validation, is also an important factor. If computer time permits, only one compound at a time should be omitted during cross validation.

4D QSAR

The 4D QSAR methodology is an extension of the 3D QSAR methodology developed by Hopfinger,[37] which considers conformational information as the fourth dimension. Similar to the CoMFA method, 4D QSAR starts off by defining a set of grid points on which molecular properties will be evaluated. In addition to the grid points, the method performs conformational ensemble sampling and uses the information obtained to evaluate grid cell occupancies. These occupancies are then used to evaluate interaction pharmacophore elements (IPEs). The IPEs together with the molecular properties are then used to develop a predictive model. When structure-based design is done to develop a quantitative model to forecast activity, the approach is termed as (RD) 3D QSAR analysis. Thus the 4D QSAR scheme can be applied to both **Receptor Dependent** (RD) and **Receptor Independent** (RI) problems. As seen earlier, (RI) 3D QSAR analysis has three inherent problems to overcome. First it is the identification of the **active** conformations/

molecular shapes of flexible compounds in the training set. In the most straightforward interpretation, particularly for *in vitro* activity, the active conformation/shape of a ligand corresponds to the receptor-bound conformation/shape. The second problem to be overcome is the specification of the basis for comparing molecules in constructing a 3D QSAR which is referred to as the **molecular alignment**. Finally, each molecule in the training set must be partitioned with respect to intermolecular (receptor) interactions. That is, different parts of each molecule can be expected to have different types of interactions with sites on a common receptor and/or in a common medium. This partitioned form of the molecule is called the **interaction pharmacophore**.

Thus the fourth dimension of 4D QSAR analysis is the 'dimension' of ensemble sampling. In 3D QSARs where each ligand molecule is represented by a single, 3D entity and is the identification of the bioactive conformation, orientation and possibly, the protonation state is a crucial step in the procedure. If the underlying pharmacophore hypothesis is based on incorrect assumptions, the resulting surrogate is hardly of any use for predictive purposes. In explicit 4D QSAR approaches, the ligands of both training and test sets are provided as an ensemble of conformations, orientations and most importantly protonation states. The most likely bioactive representation is then genetically evolved from this reservoir using a Boltzmann-weighted selection criterion.[38–41]

REFERENCES

1. Kaldor, S.W. and Kalish, V. J. *et al.* (1997). "Viracept (nelfinavir mesylate, AG1343): A potent, orally bioavailable inhibitor of HIV-1 protease." *J. Med. Chem.* 40(24): 3979–85.

2. Kempf, D.J. and Norbeck, D.W. *et al.* (1990). "Structure-based, C2 symmetric inhibitors of HIV protease." *J. Med. Chem.* 33(10): 2687–9.

3. Nagar, B. and Bornmann, W. G. *et al.* (2002). "Crystal structures of the kinase domain of c-Abl in complex with the small

molecule inhibitors PD173955 and imatinib (STI-571)." *Cancer Res.* 62(15): 4236–43.

4. Roberts, N.A. and Martin, J.A. *et al.* (1990). "Rational design of peptide-based HIV proteinase inhibitors." *Science.* 248(4953): 358–61.

5. Van de Waterbeend, H., Testa, B., Folkers, G. (eds.). (1997). *Computer-Assisted Lead Finding and Optimization.* VHCA and Wiley, Weinheim.

6. Gschwend, D.A. Good, A.C. and Kuntz, I.D. (1996). "Molecular docking towards drug discovery." *J. Mol. Recognit.* 9: 175–186.

7. Fersht, A.R., Knill Jones, J.W., Bedouelle, H. and Winter, G. (1988). "Reconstruction by site-directed mutagenesis of the transition state for the activation of tyrosine by the tyrosyl-tRNA synthetase: A mobile loop envelopes the transition state in an induced-fit mechanism." *Biochemistry.* 27: 1581–1587.

8. Richardson, J. (1868). *Medical Times and Gazette.* 2: p. 703.

9. Richet, C. and Seances, C.R. (1893). *Soc.Biol.* 9: p.775.

10. Overton, E.Z. (1897). *Physik.Chem.* 22: p.189.

11. Meyer, H. (1899). *Arch.Experim.Pathol & Pharmakoi.* 42: p.109.

12. Hammett, L.P. (1937). "The effect of structure upon the reactions of organic compounds. Benzene derivatives." *JACS.* Vol. 59. 96–98.

13. Taft, R.W. Jr. (1952). *J. Am. Chem. Soc.* 74: p.3120.

14. Hansch, C., Malony, P.P., Fujita, T., Muir, R.M. (1962). "Correlation of biological activity of phenoxyacetic acids with Hammett substituent constants with partition coefficents." *Nature.* 194: 178–180.

15. Hansch, C. and Fujita, T. (1964). "Rho-sigma-pi analysis. A method for the correlation of biological activity and chemical structure." *J. Am. Chem. Soc.* 86: 1616–1626.

16. Joshua, T. Ayers, Aaron Clauset, Jeffrey, D. Schmitt, Linda, P. Dwoskin and Peter, A. Crooks. (2005). "Molecular modeling of mono- and bis-quaternary ammonium salts as ligands at the $\alpha 4\beta$ 2* nicotinic acetylcholine receptor subtype using nonlinear technique." *The AAPS Journal.* 7 (3). Article 68.

17. Stanton, D. and Jurs, P. (1990). "Development and use of charged partial surface area structural descriptors in computer assisted quantitative structure property relationship studies." *Anal. Chem.* 62: 2323–2329.

18. Stanton, D.T., Mattioni, B.E., Knittel, J.J. and Jurs, P.C. (2004). "Development and use of hydrophobic surface area (HSA) descriptors for computer-assisted quantitative structure-activity and structure-property relationship studies." *J. Chem. Inf. Comput. Sci.* 44: 1010–1023.

19. Pimentel, G. and McClellan, A. (1960). *The Hydrogen Bond.* Reinhold Pub. Corp. New York.

20. Vinogradov, S. and Linnell, R. (1971). *Hydrogen Bonding.* Van Nostrand Reinhold, New York.

21. Palm, K., Luthman, K., Ungell, A.L., Strandlund, G., Beigi, F. and Lundahl, P.P. (1998). "A evaluation of dynamic polar molecular surface area as predictor of drug absorption: Comparison with other computational and experimental predictors." *J. Med. Chem.* 41: 5382–5392.

22. Stenberg, P., Luthman, K. and Artursson, P. (1999). "Prediction of membrane permeability to peptides from calculated dynamic molecular surface properties." *Pharm. Res.* 16: 972–978.

23. Palm, K., Stenberg, P., Luthman, K. and Artursson, P. (1997). "Polar molecular surface properties predict the intestinal absorption of drugs in humans." *Pharm. Res.* 14: 568–571.

24. Clark, D. (1999). "Rapid calculation of polar molecular surface area and its application to the prediction of transport phenomena. 2. Prediction of blood-brain barrier penetration." *J. Med. Chem.* 88: 815–821.

25. Ertl, P., Rohde, B. and Selzer, P. (2000). "Fast calculation of molecular polar surface area as a sum of fragment-based contributions and its application to the prediction of drug transport properties." *J. Med. Chem.* 43: 3714–3717.

26. Stanton, D., Mattioni, B.E., Knittel, J. and Jurs, P. (2004). "Development and use of hydrophobic surface area (HSA) descriptors for computer assisted quantitative structure-activity and structure-property relationships." *J. Chem. Inf. Comp. Sci.* 44: 1010–1023.

27. Guha, R. and Jurs, P. C. (2004). "The development of linear, ensemble and non-linear models for the prediction and interpretation of the biological activity of a set of PDGFR inhibitors." *J. Chem. Inf. Comput. Sci.* 44: 2179–2189.

28. Ferguson, A.M., Heritage, T., Jonathon, P., Pack, S.E. and Phillips, L. (1997). "EVA: A new theoretically based molecular descriptor for use in QSAR/QSPR analysis." *J. Comput. Aided Mol. Des.* 11: 143–152.

29. Myers, R.H. (1997). *Classical and Modern Regression with Applications.* PWS-KENT Publishing Company.

30. Geladi, P. and Kowalski, B.R. (1986). Partial least squares: A tutorial. *Anal. Chem. Acta.* 185: 1–17.

31. Martens, H. and Næs, T. (1989). *Multivariate Calibration.* John Wiley & Sons, New York.

32. Cramer III, R., Patterson, D. and Bunce, J. (1988). "Comparative molecular field analysis (CoMFA). I. Effect of shape on binding of steroids to carrier proteins." *J. Am. Chem. Soc.* 110: 5959–5967.

33. 1Moore, W.J. (1983). *Basic Physical Chemistry.* Prentice-Hall, London.

34. Howard, A.E. and Kollman, P.A. (1988). "An analysis of current methodologies for conformational searching of complex molecules." *J. Med. Chem.* 31: 1669–1675.

35. Saunders, M., Houk, K.N., Wu, Y.D., Still, W.C., Lipton, M., Chang, G. and Guida, W.C. (1990). "Conformations of cycloheptadecane. A comparison of methods for conformational searching." *J. Am. Chem. Soc.* 112: 1419–1427.

36. Desai, B., Sureja, D., Naliapara, Y., Shah, A. and Saxena, A. (2001). "Synthesis and QSAR studies of 4-substituted phenyl-2,6-dimethyl-3,5-bis-N-(substituted phenyl)-carbamoyl-1, 4-dihydropyridines as potential anti-tubercular agents." *Bioorg. Med. Chem.* 4: 1993–1997.

37. Prashant, S. Kharkar, Bhavik Desai, Harsukh Gaveria, Bharat Varu, Rajesh Loriya, Yogesh Naliapara, Anamik Shah and Vithal, M. Kulkarni. (2002). "Three-dimensional quantitative structure-activity relationship of 1,4-dihydropyridines as antitubercular agents." *J. Med. Chem.* 45: 4858–4867.

38. Vedani, A., Briem, H., Dobler, M., Dollinger, K. and McMasters, D.R. (2000). "Multiple conformation and protonation-state representation in 4D-QSAR: The neurokinin-1 receptor system." *J. Med. Chem.* 43: 4416–4427.

39. Vedani, A. and Dobler, M. (2000). "Multidimensional QSAR in drug research: Predicting binding affinities, toxicity, and pharmacokinetic parameters." Jucker, E. (ed.). *Progress in Drug Research*. Birkhauser Verlag. pp. 105–135.

40. Vedani, A., McMasters, D.R. and Dobler,M. (2000). "Multi-conformational ligand representation in 4D-QSAR: Reducing the bias associated with ligand alignment." *Quant. Struct. Act. Relat.* 19: 149–161.

41. Streich, D., Neuburger-Zehnder, M. and Vedani, A. (2000). "Induced fits the key for understanding LSD activity. A 4D-QSAR study on the 5-HT2A receptor system." *Quant. Struct. Act. Relat.* 19: 565–573.

Commercial Software

Accelrys (www.accelrys.com), whose Cerius2 environment includes C2.GA, C2.QSAR+, C2.CSAR, and C2.NNet; multi-Y recursive partitioning and Genetic Algorithms, Genetic Function Algorithm, Nonlinear Principal Component Analysis, and Partial Least Squares.

Tripos (www.tripos.com), which markets HQSAR and QSAR with CoMFA; and molecular field generation, PCA, PLS regression, and hierarchical clustering.

Chemical Computing Group (www.chemcomp.com), with its molecular operating environment (MOE), QuaSAR-Binary, and Binary QSAR.

Information	Source
Topological index software (MolconnZ)	http://www.edusoft-lc.com
CODESSA descriptor software	http://www.semichem.com/codessa.html
Dragon descriptor software	http://www.disat.unimib.it/chm/Dragon.htm
Pomona College QSAR site	http://clogp.pomona.edu/medchem/chem/qsar-db
CoMFA 3-D QSAR method	http://www.tripos.com/software/qsar.html
QSAR and Modelling Scoeity	http://www.ndsu.nodak.edu/qsar_soc/index.htm
Summaries of QSAR studies in literature	Journal Quantitative Structure–Activity Relationships

Recommended Books for Reading

1. *QSAR: Hansch Analysis and Related Approaches.* by Hugo Kubinyi , VCH Publishers (1993). ISBN 1-56081-768-2.

2. *3D QSAR in Drug Design: Recent Advances* (Vol 3), Edited by Hugo Kubinyi , Yvonne C. Martin, Gerd Folkers, Yvonne C. Matin, Kluwer Academic Publishers (2002). ISBN: 0-7923-4791-9.

3. *QSAR and Drug Design: New Developments and Applications*, Edited by Toshio Fujita , Elsevier (1995). ISBN-13: 978-0444886156.

4. *Comparative QSAR* by James Devillers (Editor), CRC (1998). ISBN-13: 978-1560327165.

Websites of Related Interest

1. http://michem.disat.unimib.it/chm/

2. http://www.qsarworld.com/

3. http://abagyan.scripps.edu/lab/

4. http://www.moleculardescriptors.eu/

Journals

1. *Quantitative Structure–Activity Relationships*

2. *QSAR & Combinatorial Science*

3. *Mini-Reviews in Medicinal Chemistry*

4. *SAR & QSAR in Environmental Research*

REVIEW QUESTIONS

1. Describe the principle of quantitative structure activity relationships.

2. Given below is the dataset of antifungal activity of a series of benzene derivatives against *Aspergillus niger*.

Compound	Bioactivity
1.	20%
2.	30%
3.	45%
4.	90%
5.	25%
6.	85%
7.	75%
8.	60%
9.	45%
10.	20%
11.	50%
12.	95%
13.	30%
14.	20%
15.	10%

Suggest whether a QSAR analysis can be performed on this set of compounds, justifying it.

3. For the derivatives of the following compound,

where R= $-CH_3$, $-OCH_3$, $-NH_2$, $-CH_2CH_3$, $-SH2$, $-OH$, etc. a QSAR analysis was carried out in evaluating its anticancer activity. This resulted in the following simple QSAR equation

Activity = 3.75 (Dipole moment) – 1.65 (molecular weight).

Based on this, suggest what type of chemical modification/ substitutions would enhance the bioactivity.

4. State the assumptions used in QSAR studies and some applications.

5. What are the statistical methods used in QSAR studies?

6. Explain the concept of Comparative Molecular Field Analysis (COMFA).

7. Can you suggest whether the COMFA methodology can be applied to the following set of compounds.

   Bromobenzene, bromocyclohexane, methyl bromide, ethyl bromide, bromooctane, Bromocysteine, 1-bromonaphthalene Justify your answer.

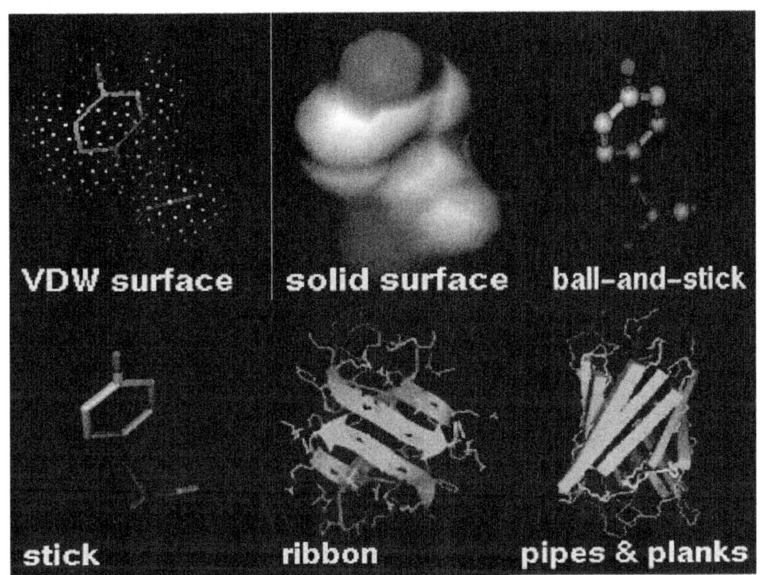

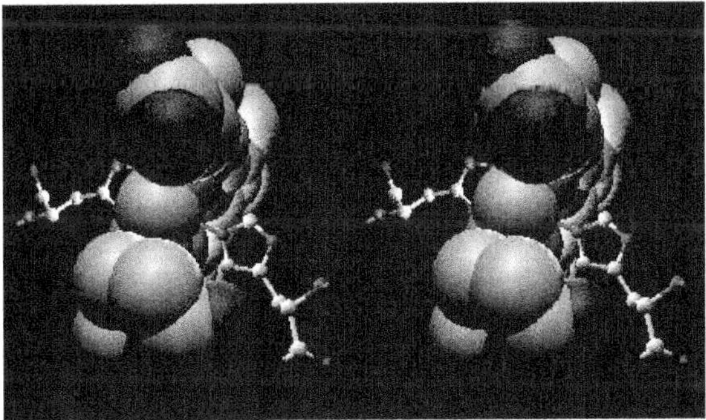

Plate 1

Plate 2

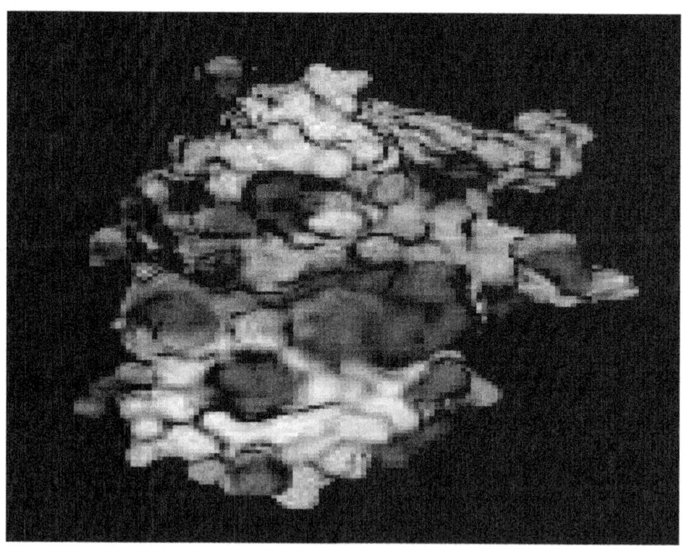

Plate 2a

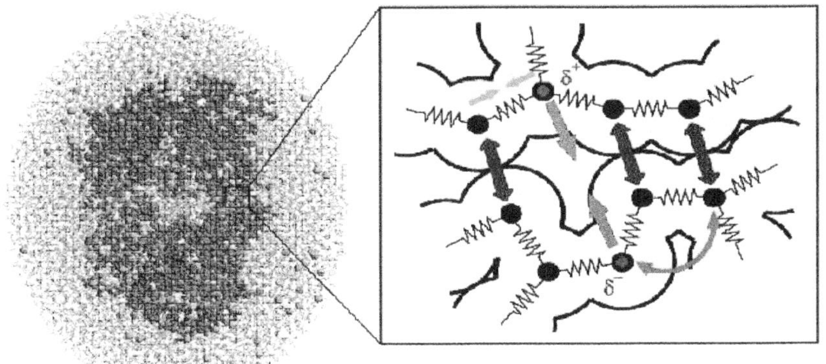

Plate 3

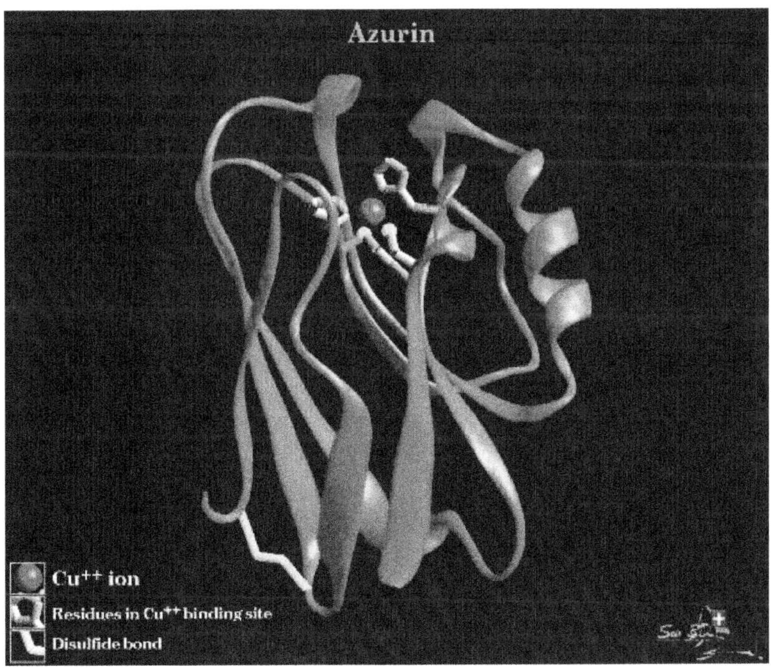

Plate 4

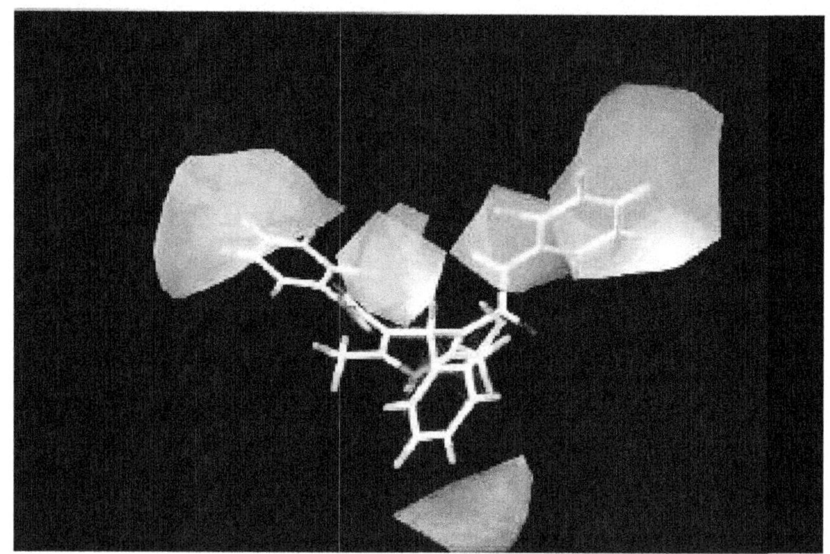

Plate 5

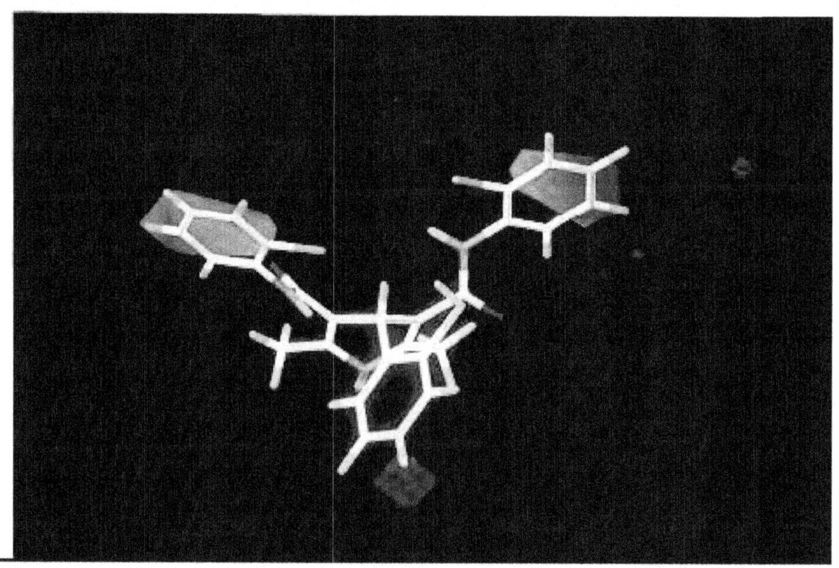

Plate 6

Glossary

Ab initio A quantum mechanical non-parametrized molecular orbital treatment (Latin: from "first principles") for the description of chemical behaviour taking into account nuclei and all electrons. In principle, it is the most accurate of the three computational methodologies: *ab initio*, semi-empirical all-valence electron methods, and molecular mechanics.

ADMET Abbreviation for the five steps in a medicine's journey through the body: absorption, distribution, metabolism, excretion and its effect, toxicology.

Agonist A molecule that triggers a cellular response by interacting with a receptor.

Angstrom A unit of length used for measuring atomic dimensions. One angstrom equals 10^{-10} metres.

Antagonist A molecule that prevents the action of other molecules, often by competing for a cellular receptor; opposite of agonist.

Anti-inflammatory A drug's ability to reduce inflammation, which can cause soreness and swelling.

Bioinformatics A field of research that relies on computers to store and analyse large amounts of biological data.

Born-Oppenheimer approximation The Born-Oppenheimer approximation consists of separating the motion of nuclei from the electronic motion. An often used physical picture is that the nuclei being so much heavier than electrons may be treated as stationary as the electrons move around them. The Schrodinger equation can then be solved for the electrons alone at a definite inter-nuclear separation.

CADD Computer-aided drug design that relies on computers, information science, statistics, mathematics, chemistry, physics, biology, and medicine.

Chemoinformatics The use of computers and algorithms to store, generate, and analyse data from combinatorial chemistry, virtual libraries, and high-throughput screening.

Clinical trial A scientific study to determine the effects of potential medicines in people; usually conducted in three phases (I, II, III), to determine whether the drug is safe, effective, and better than current therapies, respectively.

Comparative molecular field analysis Involves sampling of the steric and electrostatic field around a ligand molecule, which may provide all the information necessary for explaining its biological property in a 3D QSAR.

Connolly surface The molecular surface that is related to the solvent-accessible surface area but traced by the inward-facing part of a solvent probe model, represented by a sphere with a given radius, free to touch but not to penetrate the solute when the probe is rolled over its van der Waals surface.

Conjugate gradients A mathematical first-order procedure to minimize a function such as a potential energy function used in molecular mechanics.

CPK Corey–Pauling–Koltun or space-filling representation of a molecule in which each atom is represented by a sphere, the radius of which is proportional to the van der Waals radius of that atom.

Cyclooxygenase An enzyme, also known as COX, that makes prostaglandins from a molecule called arachidonic acid. It is the molecular target of non-steroidal anti-inflammatory drugs.

Cytochrome P450 A family of enzymes found in animals, plants, and bacteria that have an important role in drug metabolism.

De novo design A ligand design strategy in which the availability of a 3D structure of a therapeutic target (an enzyme or protein) is used to design and predict the affinity of novel ligands.

Energy minimization *In silico* method of arriving at the least energy conformation of a molecule by applying force field and minimization algorithm.

Force field A set of equations and parameters which, when evaluated for a molecular system, yields an energy as a function of the atomic coordinates.

G protein A group of switch proteins involved in a signalling system that sends incoming messages across cell membranes and within cells.

Homology modelling The art of building a protein structure knowing only its amino acid sequence and the complete 3D structure of at least one other reference protein.

Hydrogen bond A weak interaction between two electronegative atoms mediated by a hydrogen atom and that plays an important role in preserving the three-dimensional structure of protein and also in protein–ligand interaction.

Enzyme A molecule (usually a protein) that speeds up, or catalyses, a chemical reaction without being permanently altered or consumed.

Inflammation The body's characteristic reaction to infection or injury, resulting in redness, swelling, heat, and pain.

Inhibitor A molecule that "inhibits," or blocks, the biological action of another molecule.

Kinase An enzyme that adds phosphate groups to proteins.

Membrane A thin covering surrounding a cell and separating it from the environment; consists of a double layer of molecules called phospholipids and has proteins embedded in it.

Metabolism All enzyme-catalysed reactions in a living organism that builds and breaks down organic molecules, producing or consuming energy in the process.

Molecule The smallest unit of matter that retains all of the physical and chemical properties of that substance. It consists of one or more identical atoms or a group of different atoms bonded together.

Molecular descriptors Physico-chemical properties used in QSAR analysis and for computing molecular similarity.

Molecular dynamics A method for constructing the trajectory of a protein, predicting the change in conformation with parameters such as time, temperature, etc.

Neurotransmitter A chemical messenger that allows neurons (nerve cells) to communicate with each other and with other cells.

Pharmacodynamics The study of how drugs act at target sites of action in the body.

Pharmacokinetics The study of how the body absorbs, distributes, breaks down, and eliminates drugs.

Pharmacophore The spatial mutual orientation of atoms or groups of atoms assumed to be recognized by and to interact with a receptor or the active site of a receptor.

Protein A large molecule composed of one or more chains of amino acids (the building blocks of proteins) in a specific order and having a folded shape determined by the sequence of nucleotides in the gene encoding the protein. It is essential for all life processes.

Quantitative structure–activity relationships A ligand-based drug design approach wherein, from the given structures and biological data for a given set of compounds, a model is derived from which the biological activity of an untested compound can be predicted.

Receptor A specialized molecule that receives information from the environment and conveys it to other parts of the cell; the information is transmitted by a specific chemical that must fit the receptor like a key in a lock.

SMILES Simplified Molecular Input Line Entry System (SMILES) is a chemical notation system based on the principles of molecular graph theory and denotes a molecular structure as a two-dimensional graph.

Structure-based drug design An approach to developing medicines that takes advantage of the detailed, three-dimensional structure of target molecules.

Structure–activity relationship Relationship of pharmacological activity or toxicity of a xenobiotic to its chemical structure.

Toxicology The study of how poisonous substances interact with living organisms.

Virtual library A database of structures and properties of molecules that may not even have been synthesized.

X-ray crystallography A technique used to determine the detailed, three-dimensional structure of molecules based on the scattering of X-rays through a crystal of the molecule.

Index

F

File formats and molecular
   visualization 18
Flexible docking 94
Flexible ligand docking 94
Force-field approach 52
Force field scoring functions 96

G

G-protein-coupled receptors 86
Genetic function approximation
   196
GFA 196
GLIDA 87
GOLD 93
GPCRs 86

H

Hansch equation 185
HBAT 115
HEX 98
High-throughput screening 118
Hits 10
Homology modelling 5
HTS 118
Hybrid descriptors 192

I

In silico ADMET 167
Inductive learning methods 163
Ivermectin 7

K

Knowledge-based scoring
   functions 97

L

LBDD 89
Lead optimization 11, 23
Leads 10, 12
Ligand based drug design 89

Ligands 88
Lipinski's rule of five 32

M

Metabolism of drugs 170
Molecular dynamics 47,
   69, 72, 73, 74, 75
Molecular field analysis 206
   pitfalls of 206
Molecular mechanics 52
Molecular modelling 4, 5
   databases in 33
Monte Carlo simulations 5

N

Nuclear magnetic resonance
   spectroscopy 24

O

Open Babel 92

P

Partial least squares 196
PCA 195
PLS 196
Principal component analysis
   166, 195
Pubchem 156

Q

QSAR 187, 206
   evolution of 184
   methods
      types of 197
   studies
      types of descriptors in 190
      statistical methods
      used in 195
Quantitative structure activity
   relationship 183
Quantum mechanical
   simulations 48

Printed in Dunstable, United Kingdom

85055789R00138